JN006590

英語で学ぶ
制御システム設計

Introduction to Linear Control System Design

佘　錦華・禹　珍碩・大山　恭弘

【共著】

コロナ社

ま え が き

　制御（control）とは，「ある目的に適合するように，対象となっている物に所用の操作を加えること（JIS）」と定義されている。動植物を操ること，ものを操ることなどすべて「制御」であり，さまざまな手法が提案され実行され，"目的" に合う成果を上げている。本書では，電気機械システムを対象例に，対象を数式で表現し，それに基づいて目的に沿うように操る手法の基本を解説している。制御は目的通りの結果が得られたかどうかで評価されるべきものなので，マイクロコントローラを用いた実装についても述べ，実装を意識したモデリング，解析，設計のプロセスを解説した。なお，国際的に活躍する技術者となる訓練のために，本文は平易な英文で記述し，重要語には対応する日本語を併記した。"英文で学ぶ教科書" としても活用して頂きたい。章末問題は，基本問題と発展問題に分けており，発展問題は国内海外の大学院受験問題および分析的な吟味を必要とする問題を中心に選定するとともにすでに学習した知識を復習できるように問題の作成を工夫したので，理解を深める学習の一助となることを願っている。また，"教科書" は，考え方の基本が書かれているものであるが，日々新しいアプローチ法が開発されているので，読者の皆様には，この基本の考え方を実世界にどう適用し，どう修正・発展すれば "目的" を達成できるのかを考え，適応する手法を身につけて頂きたいと思う。

　最後に，本書は筆者らのこれまでの大学内外における教育経験に基づいて執筆したものである。一緒に学んだ学生・技術者の方々にこの場を借りて感謝致します。原稿を詳しくチェックし貴重な助言をくださった中国中南大学の李旻氏，中国湖南工業大学の趙凱輝氏，清水建設株式会社技術研究所の宮本皓氏，中国地質大学（武漢）の李美柳氏，周宇健氏，王澤文氏，趙大双氏と寧子昊氏に御礼を申し上げます。また，出版にあたり心暖かく見守ってくださったコロナ社に深く感謝致します。

2022 年 2 月

佘　錦華，禹　珍碩，大山　恭弘

Preface

Control is defined as *adding necessary operations to a target object to suit a certain purpose* (JIS). Manipulating animals and plants, operating things, etc. are all belong to the category of control. Various methods have been devised and implemented. Fruitful results that suited *purposes* have been achieved. In this book, we use electrical and mechanic systems as examples to explain the fundamental methods of describing systems by mathematical formulas and manipulating them according to purposes. Since control should be evaluated based on whether or not desired results are obtained, we also provide some explanations of the microcontroller-based implementation of a control system and explain implementation-conscious modeling, analysis, and design processes. To train engineers who play an active role internationally, we wrote the book in plain English and also added corresponding Japanese translations for important terms. We hope this book are used as a textbook to learn control engineering in English. Excise problems at the end of each chapter are divided into basic and advanced levels. Advanced-level problems were selected from graduate-school entrance examinations in Japan and from abroad or those requiring analytical scrutiny. We also tried to prepare excise problems in such a way that one can review the previously studied knowledge as much as possible. We hope this will play a role in helping readers deepen their understanding. In addition, a textbook mainly describes the basic way of thinking, while new approaches are being developed every day, Thus, we hope that readers think about how to apply the basic principle to the real world and how to modify and improve it to achieve your goals, and master adaptive techniques.

Finally, since this book was written based on our educational experience both inside and outside the university, we would like to take this opportunity to thank the students and engineers who studied with us. We would also like to express heartfelt appreciation to Prof. Min Li of Central South University and Prof. Kaihui Zhao of Hunan University of Technology, China; Dr. Kou Miyamoto of the Institute of Technology, Shimizu Corporation, Japan; and Dr. Meiliu Li, Mr. Yujian Zhou, Mr. Zewen Wang, Mr. Dashuang Zhao, and Mr. Zihao Ning of China University of Geosciences, Wuhan, China for their careful checking of the manuscript and for their valuable advice. We are deeply grateful to Corona Publishing Co., Ltd. for warmly watching over the publication.

February 2022

Jinhua She, Jinseok Woo, Yasuhiro Ohyama

Contents

4 | Block Diagram and Frequency Response

5 | Stability Analysis

6 | Characteristics of Control System

7 | Transfer-Function-based Control-System Design

8
Control-System Design in State Space

9
Controller Implementation

10
Comprehensive Exercises

【本書ご利用にあたって】
・本文中に記載している会社名，製品名は，それぞれ各社の商標または登録商標です。本書では Ⓡ や TM は省略しています。
・本書に記載の情報，ソフトウェア，URL は 2022 年 2 月現在のものを記載しています。
・コロナ社の Web サイトから MATLAB のサンプルデータがダウンロードできます。ぜひご利用ください。
　　https://www.coronasha.co.jp/np/isbn/9784339032383/

目　　　　次

1 | Overview of System Control

Automatic control (自動制御) plays an essential role[†1] in a wide range of fields covering engineering, physical, biological, social, and economic systems. It supports modern industry and our daily life. This chapter explains the basic concepts of *control engineering* (制御工学). We use two examples, driving a car and controlling an arm robot, to understand the necessity and importance of feedback control.

1.1 Control and Control System

Control is a series of actions applied to a plant to achieve a desired purpose[1],[†2]. According to Cambridge Dictionary, a system is a set of connected things or devices that operate together. *"A control system consists of[†3] subsystems and processes (or plants) assembled for the purpose of obtaining the desired output with desired performance, given a specified input"*[2].

We use the example of driving a car (**Figure 1.1**) to explain the basic ideas. A series of actions of starting a car is first to start the engine, then to release the brake, and finally to step on the accelerator. Ensuring the correct sequence is the key to starting a car. It is called *sequential control* (シーケンス制御). During driving, a driver adjusts the amount of

---

[†1] play a role：役割を演じる。
[†2] 肩付き数字は巻末の文献番号を示す。
[†3] consists of 〜：〜から成り立っている。

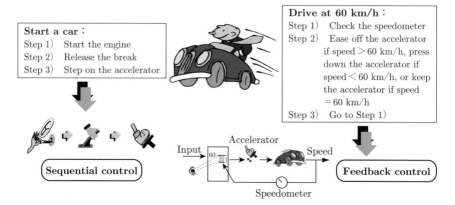

Figure 1.1   Control methods in driving a car

the accelerator based on a moving speed to keep the car moving at a target speed. Since the driver regulates the speed based on a measured speed, it is called *feedback control* (フィードバック制御).

In control engineering, a control objective is called a *plant* (制御対象), and a device that produces an adjustment for the plant is called a *controller* (制御器, コントローラ). A given command is a *reference input* (目標入力), an *output* (出力) is a controlled variable, and a *control input* (制御入力) is a signal produced by a controller.

Consider a system with the input and output being $u(t)$ and $y(t)$, respectively (**Figure 1.2**). The relationship between $u(t)$ and $y(t)$ is described as

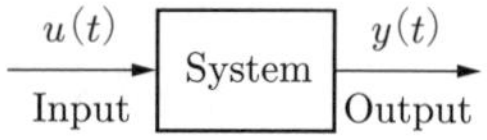

Figure 1.2   System with input and output

$$y(t) = T\{u(t)\}, \tag{1.1}$$

where T is an operator. A system is called a *static system* (静的システム) if the present output only depends on[†] the present input. For example, $y(t) = 10u(t) + u^3(t)$. And a system is called a *dynamic system* (動的システム) if

---

† depend on ～：～に依拠する，左右される。

the output of the system does not reach the steady state instantaneously
for an input. For example,

$$y(t) = \int_0^t u(\tau)\mathrm{d}\tau \text{ and } y(t) = \frac{\mathrm{d}u(t)}{\mathrm{d}t}.$$

The input-output relationship is complicated for a dynamic system. Classic
control theory uses a mathematical tool to simply describe the relationship
for a linear system in a domain other than the time (we call it the s domain).
Let it be

$$Y(s) = G(s)U(s), \tag{1.2}$$

where $U(s)$ and $Y(s)$ are the input and output signals in the s domain,
respectively; and $G(s)$ is a function in s that describes the characteristic
of the system. Note that, as an unwritten rule, we use a lower-case letter
(for example, u and y) to indicate a variable in the time domain; and an
uppercase letter (for example, U and Y), in the s domain. (1.2) shows that
the output of a system is the product of the system model and the input.
This greatly simplifies the description of a system and makes it possible to
apply algebraic tools to[†] the analysis and design of a control system.

1.2 Positioning Control of Arm Robot

Consider the problem of *positioning control* (位置決め制御) of an arm
robot on a horizontal plane. The robot has a motor, a gearbox, an arm, a
motor drive, and an encoder [**Figure 1.3**(a)]. The purpose of arm-robot
control is to ensure that the angle of the arm tracks a reference input $r(t)$
and finally stops at an angle r_0 [Figure 1.3(b)]. To make the description of
the system simple, we use a block to describe each item in a general way

---

† apply A to B : A を B に応用する (適用する, 当てはめる)。

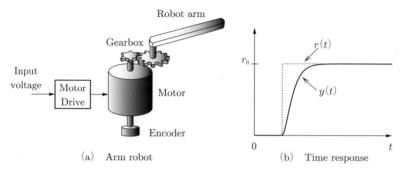

Figure 1.3   Positioning control of arm robot

(**Figure 1.4**). It is called a *block diagram* (ブロック線図). As shown in the right part in Figure 1.4, the name of an element or the input-output relationship is shown in a block, and an arrow indicates the relationship between the connected elements. The name of a variable in the time domain or the s domain is usually shown along with an arrow.

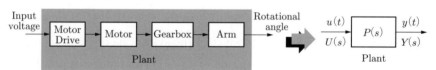

Figure 1.4   Block diagram of arm robot

We insert a controller $C(s)$ between the reference input and the control input to adjust the signal applied to the plant (**Figure 1.5**). Since

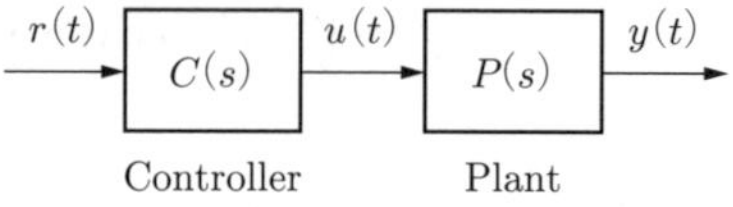

Figure 1.5   Feedforward control

$$Y(s) = P(s)U(s) = P(s)C(s)R(s), \tag{1.3}$$

a simple strategy

$$C(s) = \frac{1}{P(s)} \tag{1.4}$$

yields

$$Y(s) = R(s). \tag{1.5}$$

It ensures that the output, $y(t)$, tracks the reference input, $r(t)$. Note that signals are passed from the source to the controlled variable. The control input is generated without waiting for the effect of the source on the output. This control strategy is called *feedforward control* (フィードフォワード制御). It is simple and effective. However, it has some drawbacks. For example, the characteristics of a plant usually change with time. If $P(s)$ changes to $\hat{P}(s)$, then the output becomes

$$Y = \hat{P}CR = \hat{P}\frac{1}{P}R \neq R. \tag{1.6}$$

For simplicity, we omit the variable s in equations in the rest of this chapter unless necessary. (1.6) means that the output cannot track the reference input anymore. Moreover, there are *disturbances* (外乱) in a system. When a disturbance $d(t)$ is added to the plant (**Figure 1.6**), the output is

$$Y = PCR + PD = R + PD. \tag{1.7}$$

Clearly, the disturbance cannot be suppressed by the controller.

To solve these and other problems in feedforward control, we need to derive a new mechanism to carry out[†] system control. Con-

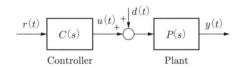

Figure 1.6   Feedforward control with disturbance

sidering that the purpose of control is to track a reference input, we first use a sensor (an encoder in this example) to measure the output (the rotational angle). Next, we use a comparator to compute the *error* (誤差):

$$e(t) = r(t) - y(t). \tag{1.8}$$

[†]   carry out ～：～を成し遂げる，実行する。

Then, we use a controller to adjust the control input according to[†1] the error. This yields a new control system structure (**Figure 1.7**), which is called *feedback control* (フィードバック制御). Note that, while feedback control uses a sensor to measure the output and uses it to generate a control input, feedforward control generates a control input without using a measured output.

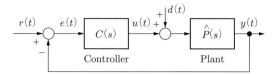

Figure 1.7   Configuration of feedback control system

Since

$$E = R - Y, \; U = C \cdot E, \; Y = \hat{P}(U + D),$$

a simple calculation gives the output of the feedback control system (Figure 1.7)

$$Y = \frac{\hat{P}C}{1 + \hat{P}C}R + \frac{\hat{P}}{1 + \hat{P}C}D. \tag{1.9}$$

If we choose the gain of the controller such that

$$|C(s)| \to \infty, \; |\hat{P}(s)C(s)| \to \infty, \tag{1.10}$$

then

$$Y = \frac{\hat{P}C}{1 + \hat{P}C}R(s) + \frac{\hat{P}}{1 + \hat{P}C}D \doteq \frac{\hat{P}C}{\hat{P}C}R + \frac{\hat{P}}{\hat{P}C}D \doteq R \tag{1.11}$$

holds even if the plant changes and there is a disturbance in the system. Thus, feedback control is used more often than feedforward control. And control usually refers to[†2] feedback control. Feedforward control is used

---

[†1]   according to ～：～にしたがって。
[†2]   refer to ～：～に言及する，～を参照する。

together with feedback control for the most part[†1].

On the other hand, a very large gain of a controller, that is[†2], $|C(s)| \to \infty$, may cause *system instability* (システムの不安定性). As we have an experience in our daily life, howling suddenly occurs (the system becomes unstable) when we turn up the volume of a speaker to the full (increase the gain of the speaker to a very big value). Thus, a design problem for feedback control is to increase the gain of a controller as much as possible while the stability of a system is guaranteed.

1.3 Classification of Systems and Control Systems

Regarding system classification, except for static and dynamic that were mentioned before, we also classify them as follows.

(1) Linear and nonlinear systems Let the relationship between the input $u(t)$ and the output $y(t)$ of a system be (1.1). For two pairs of the inputs and outputs, $\{u_1(t), y_1(t)\}$ and $\{u_2(t), y_2(t)\}$, if

$$\alpha y_1(t) + \beta y_2(t) = \mathcal{T}\{\alpha u_1(t) + \beta u_2(t)\} \tag{1.12}$$

holds for any scalars α and β, then the system is called a *linear system* (線形システム). Otherwise, it is called a *nonlinear system* (非線形システム).

(2) Continuous-time and discrete-time systems A continuous-time signal $y(t)$ is a function defined for all time t (the solid line in **Figure 1.8**). A discrete-time signal is a function defined on a sequence of time points $\{t_i\}$ $(i = 0, 1, 2, \cdots)$, that is, $y[i]$ $(i = 0, 1, 2, \cdots)$ (the circles in Figure 1.8).

A *continuous-time system* (連続時間システム) is a system in which signals

---

[†1] for the most part：大部分は，大体，ふだんは。
[†2] that is [to say]：すなわち (= namely)。

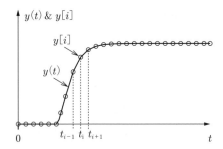

Figure 1.8   Continuous- and discrete-time signals

are continuous-time ones. On the other hand, a *discrete-time system* (離散時間システム) handles discrete-time signals. Moreover, if a system contains both continuous and discrete signals, it is called a *hybrid system* (ハイブリッドシステム).

(3) Feedback control systems   Typical feedback control systems are a *servo system* (サーボ系) (or a *serve mechanism*, サーボ機構) and a *process control system* (プロセス制御系). A servo system handles mechanical variables such as position, speed, and force. This kind of control system is widely used in robots, vehicles, and other mechatronic systems to improve motion-control performance. A process control system deals with a temperature, the amount of flow, a pressure, and a chemical variable. A *set-point control system* (定値制御システム) is usually designed for this purpose.

1.4   Key Points of Control Engineering

To perform feedback control, first, we select sensors and actuators based on control requirements. Next, we establish a model for a plant. Then, we choose and design a *control law* (制御則) for a control system. Finally, we implement the control system to carry out real-time control.

In this book, we focus on the following key issues:

i)   Modeling

ii) Analysis of system stability and characteristics

iii) Design and implementation of control laws

Modeling a plant lays the foundation of[†1] control system design. It is carried out using physical, chemical, and other natural laws. Since feedback control is a model-based control approach, a precise model of a plant not only helps us clearly understand the characteristics of the plant but also[†2] ensures good control performance.

The stability of a control system is essential. The first issue in control system design is how to *stabilize a system* (システムを安定化する). Many methods have been developed to deal with[†3] it. They provide us with analytic tools to examine the relationship between the characteristics and parameters of a control system.

After explaining the methods of *stability analysis* (安定性解析), we describe some well-used *control-system configurations* (制御システム構成) and *design methods* (設計法) for a feedback control system.

MATLAB/Simulink[†4 3)], Scilab[†5 4), 5)], and other numeric computing programs can be used to deepen the understanding of control theory and to efficiently design a control system. This book provides readers with MATLAB commands for system analysis and design.

[†1]   lay the foundation of ～：～の基礎を築く。
[†2]   not only A but also B：A だけでなく B も。
[†3]   deal with ～：～を処理する，～を扱う。
[†4]   They offer low-price licenses for students and for home use:
       https://jp.mathworks.com/
[†5]   A free, open-source software package:
       https://www.scilab.org/

──────── **Problems** ────────

⟨ **Basic Level** ⟩

[1] Draw a block diagram for a feedback control system and indicate the locations of the following elements and signals in it:

    A. controller   B. actuator   C. sensor

    D. output   E. reference input   F. error.

[2] If a control method is based on a predetermined procedure, it is called

    A. sequential control   B. feedback control.

[3] Choose dynamic systems from the following list:

    A. a swinging pendulum   B. a gearbox

    C. a bouncing ball   D. a vehicle

    E. a wardrobe   F. an electrical switch.

[4] A comparison between an output and a reference input is done by

    A. a comparator   B. a controller

    C. a feedback element   D. a sensor.

[5] Choose a suitable combination below to fill in the blanks:

(α) applies an operation to a plant to (β) a/an (γ) when a/an (γ) occurs.

    A. α: sequential control, β: reduce, γ: correction

    B. α: sequential control, β: increase, γ: error

    C. α: feedback control, β: reduce, γ: error

    D. α: feedback control, β: increase, γ: correction.

[6] Classify the following control systems into two kinds: sequential control and feedback control:

    A. the speed control of a vehicle

    B. the temperature control of a refrigerator

    C. an electronic calculator

    D. a traffic light

    E. a coffee vending machine.

[7] Choose the advantages of a feedforward control system:

    A. It is economical and easy to maintain.

    B. The calibration is easy.

C. The system is simple and the number of components is smaller than a feedback system.

D. It is robust to a disturbance.

【8】 Draw a block diagram for a room-temperature-control system.

【9】 Draw a block diagram for the door-control system of a stockroom (**Figure 1.9**).

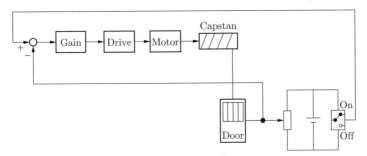

Figure 1.9 Problem 9

Advanced Level

【1】 Explain a control sequence for a rice cooker.

【2】 List basic requirements for the auto-focus system of a camera.

2 Mathematical Foundations of Control Theory

We use control theory to analyze and synthesize a control system based on a mathematical model. This chapter provides a short review of engineering mathematics that is necessary for modeling and analysis in control theory. Refer to college textbooks for calculus.

2.1 Complex Number

A *complex number* (複素数) is indicated by

$$z = a + jb \quad \text{[Orthogonal form, \textbf{Figure 2.1}(a)]}, \quad (2.1)$$

$$= re^{j\theta} \quad \text{[Polar form, Figure 2.1(b)]}, \quad (2.2)$$

where a, b, r, and θ are real numbers and $j \, (= \sqrt{-1})$ is the imaginary unit[†].

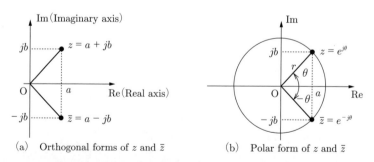

(a) Orthogonal forms of z and $\bar{z}$ (b) Polar form of z and $\bar{z}$

Figure 2.1 Complex number and complex conjugate

[†] Note that the imaginary unit is indicated by i in high-school mathematics. However, since i is used for current in electrical engineering, j is used instead to avoid confusion. Control engineering inherits this tradition.

a and b in (2.1) are the real and imaginary parts of z, respectively; and r and θ in (2.2) are the distance of the point z in the complex plane to the origin and the angle of the line Oz with the positive real axis, respectively (Figure 2.1).

Euler's formula (オイラーの公式)

$$e^{j\theta} = \cos\theta + j\sin\theta \tag{2.3}$$

provides us with the relationships between a, b, r, and θ:

$$a = r\cos\theta, \ b = r\sin\theta, \ r = \sqrt{a^2 + b^2}, \ \theta = \tan^{-1}\frac{b}{a}. \tag{2.4}$$

e in (2.3) is a constant, which is called Euler's number or Napier's constant. It is the base of the natural logarithm. The definition of e is given by

$$e \ = \ \lim_{n\to\infty}\left(1 + \frac{1}{n}\right)^n = \lim_{x\to 0}(1 + x)^{1/x} = 2.71828\cdots \tag{2.5}$$

The *complex conjugate* (共役複素数) of z is $\bar{z} = a - jb$. The only difference between z and $\bar{z}$ is the sign of the imaginary part. Or in other words, z and $\bar{z}$ are symmetric with respect to the real axis.

Let a_1, a_2, b_1, b_2, r_1, r_2, θ_1, and θ_2 be real numbers, and $z_1 \ (= a_1 + jb_1 = r_1 e^{j\theta_1})$ and $z_2 \ (= a_2 + jb_2 = r_2 e^{j\theta_2})$ be two complex numbers. The *four arithmetic operations* (四則演算) for z_1 and z_2 are defined to be

Addition (足し算):

$$z_1 + z_2 = (a_1 + a_2) + j(b_1 + b_2) = r_1 e^{j\theta_1} + r_2 e^{j\theta_2}; \tag{2.6}$$

Subtraction (引き算):

$$z_1 - z_2 = (a_1 - a_2) + j(b_1 - b_2) = r_1 e^{j\theta_1} - r_2 e^{j\theta_2}; \tag{2.7}$$

Multiplication (掛け算):

$$z_1 \times z_2 = (a_1 + jb_1) \times (a_2 + jb_2) = (a_1 a_2 - b_1 b_2) + j(a_1 b_2 + a_2 b_1)$$

$$= r_1 e^{j\theta_1} \times r_2 e^{j\theta_2} = (r_1 r_2) e^{j(\theta_1 + \theta_2)}; \tag{2.8}$$

Division (割り算) $(a_2 + jb_2 \neq 0$ and $r_2 \neq 0)$:

$$z_1 \div z_2 = \frac{a_1 + jb_1}{a_2 + jb_2}$$

$$= \frac{(a_1 a_2 + b_1 b_2) + j(a_2 b_1 - a_1 b_2)}{a_2^2 + b_2^2}$$

$$= r_1 e^{j\theta_1} \div (r_2 e^{j\theta_2})$$

$$= \frac{r_1}{r_2} e^{j(\theta_1 - \theta_2)}. \tag{2.9}$$

───── *de Moivre's formula* (ド・モアブルの定理) ─────

$$(\cos\theta + j\sin\theta)^n = \cos n\theta + j\sin n\theta \tag{2.10}$$

holds for a real number θ and an integer n.

──────────────────────────────

【**Example 2.1**】   Calculate the polar form of $z = \sqrt{12} + j2$.

Since

$$r = |z| = \sqrt{(\sqrt{12})^2 + 2^2} = 4 \text{ and}$$

$$\theta = \tan^{-1}\frac{2}{\sqrt{12}} = \tan^{-1}\frac{\sqrt{3}}{3} = \frac{\pi}{6}, \ z = 4e^{j\pi/6}.$$

──────────────────────────────

Let $z_0 = x_0 + jy_0$, $z_1 = x_1 + jy_1$, $z = x + jy$, and a rotational angle be α. The complex-number forms of coordinate translations are

Parallel translation: $z = z_1 + z_0$, $\tag{2.11}$

Rotational translation: $z = z_0 e^{j\alpha}$. $\tag{2.12}$

And

Parallel translation $\begin{cases} x = x_1 + x_0, \\ y = y_1 + y_0; \end{cases}$ $\tag{2.13}$

Rotational translation $\begin{cases} x = x_0 \cos\alpha - y_0 \sin\alpha, \\ y = x_0 \sin\alpha + y_0 \cos\alpha. \end{cases}$ (2.14)

2.2 Trigonometric Functions

Before explaining *trigonometric functions* (三角関数), we first explain the concept of *radians* (ラジアン). An angle is commonly measured in degrees. However, calculus and mathematical analysis use the unit *radian* for an angle, which is the *SI unit* (国際単位, SI 単位) of angular measure. As mentioned in the SI brochure, while[†] *degree* is not an SI unit, it is an accepted unit.

One radian is defined to be the angle that subtends the arc of length equal to the radius of the circle (**Figure 2.2**). Note that the arc length l is $2\pi r$ for $360°$, that is, $\theta = 2\pi r/r = 2\pi$ rad for $360°$. Thus,

$$1\,\text{rad} = \frac{360°}{2\pi} = \frac{180°}{\pi}.$$ (2.15)

We can use a unit circle to define trigonometric functions (**Figure 2.3**)

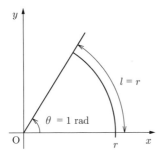

Figure 2.2 Definition of one radian

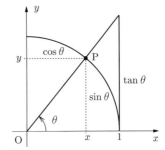

Figure 2.3 Definition of trigonometric functions

[†] while ～, ⋯ : ～, しかるに（一方では）⋯, but とほぼ同じ意味・用法。

$$\cos\theta = \frac{x}{1} = x, \ \ \sin\theta = \frac{y}{1} = y, \ \ \tan\theta = \frac{\sin\theta}{\cos\theta} = \frac{y}{x}. \tag{2.16}$$

Note that $\sin\theta/\tan\theta = x/1$ holds because of the similarity between the related triangles in Figure 2.3. It gives $\tan\theta$ in (2.16).

It is clear from Euler's formula that

$$\cos\theta = \frac{e^{j\theta} + e^{-j\theta}}{2}, \ \ \sin\theta = \frac{e^{j\theta} - e^{-j\theta}}{2j}. \tag{2.17}$$

Trigonometric addition formulas (三角関数の加法定理) are

$$\sin(\alpha \pm \beta) = \sin\alpha\cos\beta \pm \cos\alpha\sin\beta, \tag{2.18}$$

$$\cos(\alpha \pm \beta) = \cos\alpha\cos\beta \mp \sin\alpha\sin\beta, \tag{2.19}$$

$$\tan(\alpha \pm \beta) = \frac{\tan\alpha \pm \tan\beta}{1 \mp \tan\alpha\tan\beta}. \tag{2.20}$$

Consider the sum of the form $a\sin\theta + b\cos\theta$ $(a, b \neq 0)$. We use a and b to construct a right-angled triangle (**Figure 2.4**). Thus,

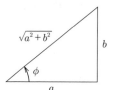

Figure 2.4   Construction of right-angled triangle

$$\sin\phi = \frac{b}{\sqrt{a^2 + b^2}}, \ \ \cos\phi = \frac{a}{\sqrt{a^2 + b^2}}. \tag{2.21}$$

As a result,

$$a\sin\theta + b\cos\theta = \sqrt{a^2 + b^2}\left(\frac{a}{\sqrt{a^2 + b^2}}\sin\theta + \frac{b}{\sqrt{a^2 + b^2}}\cos\theta\right)$$

$$= \sqrt{a^2 + b^2}\left(\cos\phi\sin\theta + \sin\phi\cos\theta\right)$$

$$= \sqrt{a^2 + b^2}\sin(\theta + \phi), \tag{2.22}$$

$$\phi = \tan^{-1}\frac{b}{a}. \tag{2.23}$$

2.3   Laplace Transform

The *Laplace transform* (ラプラス変換) is a tool for solving ordinary differential equations. It transforms complex differential equations into simple algebraic ones. Solving those algebraic equations and carrying out the *inverse Laplace transforms* (ラプラス逆変換) allow us to find the solutions of ordinary differential equations in an easy way (**Figure 2.5**).

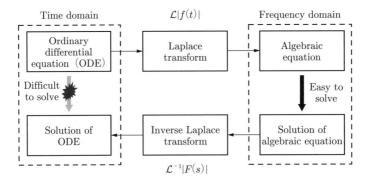

Figure 2.5   Calculation flow of using Laplace
transform to solve differential equations

Since *dynamic models* (動的モデル) of systems are usually differential equations, it is difficult to directly solve them. To avoid this difficulty, classical control theory uses the Laplace transform to model, analysis, and synthesize a control system.

The Laplace transform of a time-domain function $f(t)$ $(t \geq 0)$[†] exists if $f(t)$ is piecewise continuous for $0 \leq t < \infty$ and is of exponential order (that is, there exists a real number σ such that $\lim_{t \to \infty} e^{\sigma t}|f(t)| = 0$.). It is defined to be

---

[†]   We usually set the start time for control to be $t = 0$ and only pay attention to $t \geq 0$.

$$F(s) = \int_0^\infty f(t)e^{-st}\mathrm{d}t, \tag{2.24}$$

where $s = \sigma + j\omega$ is a complex variable. It is indicated by

$$F(s) = \mathcal{L}\{f(t)\}. \tag{2.25}$$

The complex variable s has two parts: real and imaginary. The real part σ corresponds to an exponential term $(e^{\sigma t})$; and the imaginary part ω†, a sine function in the time domain. Our primary concern is a steady-state response to a sine signal, which can be evaluated by setting $s = j\omega$. Note that $\omega = 0$ gives the DC (*direct current,* 直流) response. On the other hand, the relationship between an angular frequency ω [rad/s] and a frequency f [Hz] is $\omega = 2\pi f$. The Laplace transforms of some typical functions (**Table 2.1**) are calculated based on the definition.

Table 2.1 Laplace transforms of typical functions

$f(t)$	$F(s)$	$f(t)$	$F(s)$
$\delta(t)$	1	t^n	$\dfrac{n!}{s^{n+1}}$
$1(t)$	$\dfrac{1}{s}$	$t^n e^{-at}$	$\dfrac{n!}{(s+a)^{n+1}}$
t	$\dfrac{1}{s^2}$	$e^{-at}\sin\omega t$	$\dfrac{\omega}{(s+a)^2+\omega^2}$
e^{-at}	$\dfrac{1}{s+a}$	$e^{-at}\cos\omega t$	$\dfrac{s+a}{(s+a)^2+\omega^2}$
$\sin\omega t$	$\dfrac{\omega}{s^2+\omega^2}$	$\sinh\omega t$	$\dfrac{\omega}{s^2-\omega^2}$
$\cos\omega t$	$\dfrac{s}{s^2+\omega^2}$	$\cosh\omega t$	$\dfrac{s}{s^2-\omega^2}$

$\delta(t)$ in Table 2.1 is the *Dirac delta function* (ディラックのデルタ関数) and also called the *unit impulse function* (単位インパルス関数). Let

$$\delta_\epsilon(t) = \begin{cases} \dfrac{1}{\epsilon}, & 0 < t \leqq \epsilon, \\ 0, & \text{otherwise.} \end{cases} \tag{2.26}$$

† "," の使用で同じ動詞の繰返しを避ける：このコンマは "corresponds to" を意味する。

$\delta(t)$ is defined to be $\delta(t) = \lim_{\epsilon \to 0} \delta_\epsilon(t)$ (**Figure 2.6**). It is used to describe a shock and has the following characteristics:

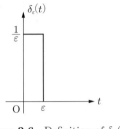

Figure 2.6 Definition of $\delta_\epsilon(t)$

$$\begin{cases} \int_0^\infty \delta(t)\mathrm{d}t = 1, \\ \int_0^\infty f(t)\delta(t-a)\mathrm{d}t = f(a). \end{cases} \quad (2.27)$$

The *unit step function* (単位ステップ関数), which is defined to be

$$1(t) = \begin{cases} 0, & t < 0, \\ 1, & t \geq 0, \end{cases} \quad (2.28)$$

is widely used as a reference signal to analyze system characteristics and verify the performance of a designed control system.

The Laplace transform has many properties that make it easy to analyze a linear dynamical system. Some properties are listed in **Table 2.2**, where

$$F_i(s) = \mathcal{L}\{f_i(t)\}, \ i = 1,2; \ f(t) * g(t) = \int_0^t f(\tau)g(t-\tau)\mathrm{d}\tau; \ (2.29)$$

and a and b are constants. We assume that $f(t) = 0$ for $t < 0$. A point to be aware of[1] using the final-value theorem is that the calculation result is correct only when the final value exists. So, a careful examination of the existence of the final value is needed before we use the theorem. For example, $f(\infty) \neq 0$ for $f(t) = \sin t$, but $\lim_{s \to 0} sF(s) = 0$.

The differentiation and integration properties are of[2] great importance in control engineering. Their proofs are shown below.

$$\mathcal{L}\left\{\frac{\mathrm{d}f(t)}{\mathrm{d}t}\right\} = \int_0^\infty \frac{\mathrm{d}f(t)}{\mathrm{d}t}e^{-st}\mathrm{d}t = \int_0^\infty e^{-st}\mathrm{d}f(t)$$
$$= e^{-st}f(t)\big|_0^\infty - \int_0^\infty f(t)\mathrm{d}e^{-st} = -f(0) + \int_0^\infty sf(t)e^{-st}\mathrm{d}t$$

---

[1] be aware of $\sim$：〜を認識する，把握している。
[2] of + 抽象名詞 = 形容詞。

Table 2.2 Properties of Laplace transform

Name	Formula
Linearity	$\mathcal{L}\{af_1(t) + bf_2(t)\} = aF_1(s) + bF_2(s)$
Differentiation	$\mathcal{L}\left\{\dfrac{\mathrm{d}f(t)}{\mathrm{d}t}\right\} = sF(s) - f(0)$ $\mathcal{L}\dfrac{\mathrm{d}^n f(t)}{\mathrm{d}t^n} = s^n F(s) - s^{n-1} f(0)$ $\left. - s^{n-2}\dfrac{\mathrm{d}f(t)}{\mathrm{d}t}\right|_{t=0} - \cdots - \left.\dfrac{\mathrm{d}^{(n-1)}f(t)}{\mathrm{d}t^{(n-1)}}\right|_{t=0}$ $\mathcal{L}\{tf(t)\} = -\dfrac{\mathrm{d}F(s)}{\mathrm{d}s}$ $\mathcal{L}\{t^n f(t)\} = (-1)^n F^{(n)}(s)$ $\mathcal{L}\left\{t^{-1}f(t)\right\} = \displaystyle\int_s^\infty F(\sigma)\mathrm{d}\sigma$
Integration	$\mathcal{L}\left\{\displaystyle\int_0^t f(\tau)\mathrm{d}\tau\right\} = \dfrac{1}{s}F(s)$ $\mathcal{L}\left\{\underbrace{\displaystyle\int_0^t \mathrm{d}t \int_0^t \mathrm{d}t \cdots \int_0^t f(t)\mathrm{d}t}_{n}\right\} = \dfrac{1}{s^n}F(s)$
Convolution	$\mathcal{L}\{f_1(t) * f_2(t)\} = F_1(s)F_2(s)$
Time shifting	$\mathcal{L}\{f(t - T)\} = e^{-sT}F(s)$
Frequency shifting	$\mathcal{L}\{e^{-at}f(t)\} = F(s + a)$
Time scaling	$\mathcal{L}\{f(at)\} = \dfrac{1}{a}F\left(\dfrac{s}{a}\right)$
Initial value	$f(0_+) = \lim\limits_{s\to\infty} sF(s)$
Final value	$f(\infty) = \lim\limits_{s\to 0} sF(s)$

$$= s\left[\int_0^\infty f(t)e^{-st}\mathrm{d}t\right] - f(0) = sF(s) - f(0). \tag{2.30}$$

Note that, if $f(t)$ is discontinuous at $t = 0$, we choose $t = 0$ from the positive side, that is, $f(0_+)$.

On the other hand, let $g(t) = \displaystyle\int_0^t f(\tau)\mathrm{d}\tau$. Applying the differentiation property to $g(t)$ yields

$$\mathcal{L}\left\{\frac{\mathrm{d}g(t)}{\mathrm{d}t}\right\} = s\mathcal{L}\{g(t)\} - g(0). \tag{2.31}$$

Since $g(0) = 0$, $\mathcal{L}\{f(t)\} = s\mathcal{L}\{g(t)\}$. Therefore,

$$\mathcal{L}\left\{\int_0^t f(\tau)\mathrm{d}\tau\right\} = \frac{F(s)}{s}. \tag{2.32}$$

[Example 2.2] Calculate the Laplace transform of $f(t) = \sin 2t \sin 3t$.

Note that

$$f(t) = \sin 2t \sin 3t = \frac{\cos(2t-3t) - \cos(2t+3t)}{2} = \frac{\cos t - \cos 5t}{2}.$$

From Table 2.1, we have

$$F(s) = \mathcal{L}\{f(t)\} = \frac{\mathcal{L}\{\cos t\} - \mathcal{L}\{\cos 5t\}}{2} = \frac{1}{2}\left(\frac{s}{s^2+1} - \frac{s}{s^2+5^2}\right)$$

$$= \frac{12s}{(s^2+1)(s^2+25)}.$$

The following MATLAB commands yield the calculation result:

```
      Program 2.1   Laplace transform
1  syms t; % A symbolic variable
2  f1=sin(2*t) % f1(t)
3  f2=sin(3*t) % f2(t)
4  F=laplace(f1*f2) % Laplace transform
```

The definition of the *inverse Laplace transform* (ラプラス逆変換) is

$$f(t) = \frac{1}{2\pi j}\int_{\sigma-j\infty}^{\sigma+j\infty} F(s)e^{st}\mathrm{d}s \tag{2.33}$$

and is indicated by

$$f(t) = \mathcal{L}^{-1}\{F(s)\}. \tag{2.34}$$

Since $F(s)$ is a rational function of s, we can use the partial-fraction expansion and the properties in Table 2.2 and look up Table 2.1 to find inverse Laplace transforms.

When $F(s)$ has only nonrepeated poles ($p_i \neq p_j$ for $i \neq j$ and $i, j = 1, 2, \cdots, n$)

$$F(s) = \frac{N(s)}{D(s)} = \frac{K(s+z_1)(s+z_2)\cdots(s+z_m)}{(s+p_1)(s+p_2)\cdots(s+p_n)}, \quad m \leq n, \tag{2.35}$$

it can be expanded into a sum of simple partial fractions

$$F(s) = \frac{A_1}{s + p_1} + \frac{A_2}{s + p_2} + \cdots + \frac{A_n}{s + p_n}, \tag{2.36}$$

$$A_i = (s + p_i)\frac{N(s)}{D(s)}\bigg|_{s = -p_i}. \tag{2.37}$$

Since $\mathcal{L}^{-1}\left\{\dfrac{A_i}{s + p_i}\right\} = A_i e^{-p_i t}$,

$$f(t) = \mathcal{L}^{-1}\{F(s)\} = A_1 e^{-p_1 t} + A_2 e^{-p_2 t} + \cdots + A_n e^{-p_n t}. \tag{2.38}$$

When $F(s)$ has repeated poles†,

$$F(s) = \frac{N(s)}{D(s)} = \frac{K(s + z_1)(s + z_2)\cdots(s + z_m)}{(s + p)^n}, \quad m \le n, \tag{2.39}$$

$$f(t) = \operatorname*{Res}_{s = -p}\left[\frac{N(s)}{D(s)}e^{st}\right]$$

$$= \frac{1}{(n-1)!}\lim_{s \to -p}\frac{\mathrm{d}^{n-1}}{\mathrm{d}s^{n-1}}\left[(s + p)^n \frac{N(s)}{D(s)}e^{st}\right]. \tag{2.40}$$

【**Example 2.3**】 Calculate the inverse Laplace transform of

$$F(s) = \frac{1}{s(s + 1)(s + 2)}.$$

Let

$$F(s) = \frac{1}{s(s + 1)(s + 2)} = \frac{A_1}{s} + \frac{A_2}{s + 1} + \frac{A_3}{s + 2}.$$

Then,

$$A_1 = F(s)s|_{s=0} = \frac{1}{2}, \quad A_2 = F(s)(s + 1)|_{s=-1} = -1,$$

$$A_3 = F(s)(s + 2)|_{s=-2} = \frac{1}{2}.$$

Thus,

$$F(s) = \frac{1}{2}\frac{1}{s} - \frac{1}{s + 1} + \frac{1}{2}\frac{1}{s + 2}.$$

---

† The residue theorem is used to carry out the calculation in (2.37) and (2.40).

We have

$$f(t) = \mathcal{L}^{-1}\{F(s)\} = \frac{1}{2} - e^{-t} + \frac{1}{2}e^{-2t}.$$

The MATLAB commands are as follows:

Program 2.2   Inverse Laplace transform

```
1  syms s; % A symbolic variable
2  F=1/(s*(s+1)*(s+2)) % F(s)
3  f=ilaplace(F) % Inverse Laplace transform
```

2.4   Logarithmic Graph

A *semi-logarithmic graph* (片対数グラフ) is a graph in which the scales on the x- and y-axes are not evenly spaced but one of them has logarithmic intervals. The value of $\log_{10} x$ increases by 1 for every 10 times increase of x (**Table 2.3**).

The scale of a logarithmic[†] graph is shown in **Figure 2.7**.

Table 2.3   Logarithmic function

x	0.01	0.1	1	10	100	10^n
$\log_{10} x$	-2	-1	0	1	2	n

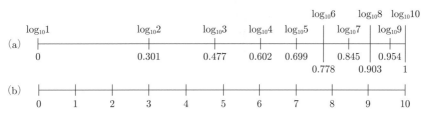

Figure 2.7   (a) Logarithmic scale and (b) linear scale

---

† Common logarithm：常用対数, natural logarithm：自然対数。

2.5   Unit of Gain

Decibel (デシベル) is a unit of ratio expressed using logarithms to compare the magnitudes of quantities, such as voltage, current, power, and volume. It is indicated by dB. We use

$$L_p = 10 \log_{10} \frac{P}{P_{ref}} \ [\text{dB}] \tag{2.41}$$

for a *power gain* (電力利得) (P: power, P_{ref}: the reference value of P). On the other hand, since power is typically proportional to the square of voltage or current, we use

$$L_v = 10 \log_{10} \frac{V^2}{V_{ref}^2} = 10 \log_{10} \left(\frac{V}{V_{ref}} \right)^2 = 20 \log_{10} \frac{V}{V_{ref}} \ [\text{dB}] \tag{2.42}$$

for a *voltage gain* (電圧利得) (V: power, V_{ref}: the reference value of V). The same is true for a current gain.

—————— **Problems** ——————

<一 **Basic Level** 一>

[1] Calculate the Cartesian coordinates of the orthogonal form, the polar coordinates, and the conjugate numbers of the following complex numbers:

    A. $\dfrac{1}{3 + j2}$ B. $\dfrac{(3 + j4)(2 - j5)}{j2}$ C. $1 + j\sqrt{3}$ D. $\dfrac{j2}{-1 + j}$.

[2] Calculate the lengths and angles for $z_1 = \sqrt{3} + j$ and $z_2 = \dfrac{3}{2} + j\dfrac{3\sqrt{3}}{2}$.

[3] Calculate the following formulas for $z_1 = \sqrt{3} + j$ and $z_2 = \dfrac{3}{2} + j\dfrac{3\sqrt{3}}{2}$:

    A. $z_1 + z_2$ B. $z_1 - z_2$ C. $z_1 \times z_2$ D. $z_1 \div z_2$.

[4] Convert the following angles from degrees to radians, and vice versa:

   A. 30°   B. 45°   C. 60°

   D. 2 rad   E. 1.5 rad   F. 0.5 rad.

[5] Calculate the values of the following trigonometric functions:

   A. $\sin 120°$ B. $\tan 210°$

   C. $\sin 75° - \sin 15°$ D. $\tan 210° - \cos 330°$.

[6] Find the period of $f(t) = (10 \cos t)^2$.

[7] Find the Laplace transforms $F(s)$ of the time functions $f(t)$.

   A. $f(t) = 6 \times 1(t) + 4e^{-t}$ B. $f(t) = \delta(t) - ae^{-at}$

   C. $f(t) = e^{at} \sin \omega t$ D. $f(t) = t^3 e^{-5t}$.

[8] Find the time functions $f(t)$ of Laplace transforms $F(s)$.

   A. $F(s) = \dfrac{s}{s^2 + \omega^2} e^{-3s}$ B. $F(s) = \dfrac{3s + 4}{s^2 + 3s + 2}$

   C. $F(s) = \dfrac{3}{(s+2)(s+5)}$ D. $F(s) = \dfrac{4s+1}{s^2 - 16}$.

[9] Find the Laplace transforms of the following functions and use MATLAB to verify the results:

   A. $f(t) = e^{-bt} \sin at$ B. $f(t) = \sin^2 \omega t$.

[10] Find the Laplace transform of $f(t) = \begin{cases} 1 - t, & 0 \le t \le 1, \\ 0, & t > 1. \end{cases}$

[11] Find the inverse Laplace transforms of the following functions and use MATLAB to verify the results:

   A. $F(s) = \dfrac{1}{(s+1)(s-2)(s+3)}$ B. $F(s) = \dfrac{a}{s^2 + \omega^2}$.

[12] What is the gain in decibel when an input voltage is 100 V and an output voltage is 200 V?

   A. 2   B. 3.01   C. 6.02   D. 12.04.

Advanced Level

[1] Let $f(t)$ be a periodic function with its period being $2b$

$$f(t) = \begin{cases} t, & 0 \le t < b, \\ 2b - t, & b \le t < 2b, \end{cases} \quad \text{and } f(t + 2b) = f(t).$$

Calculate its Laplace transform.

[2] Calculate the inverse Laplace transform of $F(s) = 1/s^2(s+1)$ and use MATLAB to verify the result.

3

Modeling of Dynamic Systems

As explained in Chapter 1, systems are divided into static and dynamic:

(1) Static system   The output of a system depends only on the current input (for example, a voltage divider and a combinational logic circuit). The input-output relationship of such a system is described using a simple algebraic equation.

(2) Dynamic system   The output of a system depends on past and/or future input (for example, an integrator, a differentiator, and an arm robot). The input-output relationship of such a system is described by a differential equation.

Modeling (モデリング) is the process of generating a mathematical model to describe the input-output relationship of a system. There are two kinds of models: models based on experiments and those based on laws of nature (for example, physical, chemical, and biological laws). We focus on models derived using the most common physical laws in this introductory text.

In this chapter, we first use basic elements in mechanical and electrical engineering to explain models in the time and the s domains. Then, we show how to build a linear approximate model for a nonlinear system.

3.1   State-Space Representation

A model of a linear system is usually described by a high-order ordinary differential equation. Nevertheless, we can describe a system using first-order differential equations, which is called a *state-space representation* (状態空間表現). The state-space representation of a linear, time-invariant, finite-dimensional system is

$$\begin{cases} \dot{x}(t) = Ax(t) + Bu(t), \\ y(t) = Cx(t) + Du(t), \end{cases} \tag{3.1}$$

where $x(t)$ $(\in \mathbb{R}^n)^\dagger$ is a state vector, $u(t)$ $(\in \mathbb{R}^r)$ is a control input vector, and $y(t)$ $(\in \mathbb{R}^m)$ is an output vector, A $(\in \mathbb{R}^{n \times n})$ is a system matrix, B $(\in \mathbb{R}^{n \times r})$ is an input matrix, C $(\in \mathbb{R}^{m \times n})$ is an output matrix, and D $(\in \mathbb{R}^{m \times r})$ is a direct transmission term.

This section uses three examples to show how to build a state-space model from an ordinary differential equation.

【Example 3.1】 Consider a spring-mass-damper system (**Figure 3.1**), where K [N/m] is an elastic modulus, M [kg] is a mass, and b [N · s/m] is a viscous-friction coefficient. The equation of motion is

$$M\frac{\mathrm{d}^2 y(t)}{\mathrm{d}t^2} = f(t) - Ky(t) - b\frac{\mathrm{d}y(t)}{\mathrm{d}t}. \tag{3.2}$$

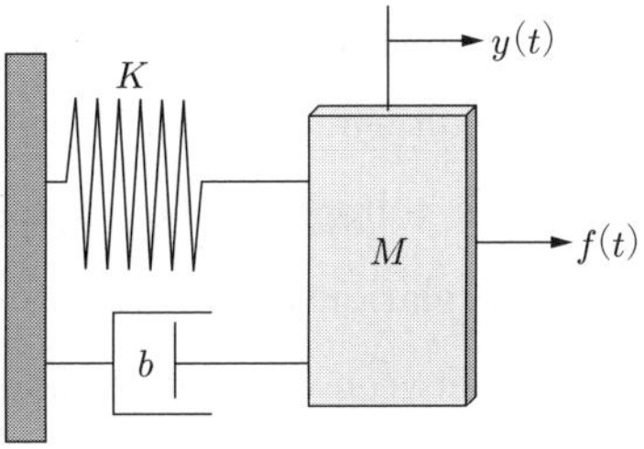

Figure 3.1 Spring-mass-damper system

If we choose the displacement and the speed of the mass to be the state, that is, $x(t) = [x_1(t),\ x_2(t)]^\mathrm{T} = [y(t),\ \dot{y}(t)]^\mathrm{T}$, then we can write (3.2) to be

$$\begin{cases} \dfrac{\mathrm{d}x_1(t)}{\mathrm{d}t} = x_2(t), \\ \dfrac{\mathrm{d}x_2(t)}{\mathrm{d}t} = -\dfrac{K}{M}x_1(t) - \dfrac{b}{M}x_2(t) + \dfrac{1}{M}f(t), \\ y(t) = x_1(t), \end{cases} \tag{3.3}$$

$\dagger$ $\mathbb{R}^n$ is a real space of dimension n and $\mathbb{R}^{n \times m}$ is a set of $n \times m$ real matrices.

or

$$\begin{cases} \dfrac{\mathrm{d}}{\mathrm{d}t}\begin{bmatrix} x_1(t) \\ x_2(t) \end{bmatrix} = \begin{bmatrix} 0 & 1 \\ -\dfrac{K}{M} & -\dfrac{b}{M} \end{bmatrix}\begin{bmatrix} x_1(t) \\ x_2(t) \end{bmatrix} + \begin{bmatrix} 0 \\ \dfrac{1}{M} \end{bmatrix} u(t), \\[2em] y(t) = \begin{bmatrix} 1 & 0 \end{bmatrix}\begin{bmatrix} x_1(t) \\ x_2(t) \end{bmatrix} + \begin{bmatrix} 0 \end{bmatrix} u(t). \end{cases} \tag{3.4}$$

Thus, we have

$$A = \begin{bmatrix} 0 & 1 \\ -\dfrac{K}{M} & -\dfrac{b}{M} \end{bmatrix}, \ B = \begin{bmatrix} 0 \\ \dfrac{1}{M} \end{bmatrix}, \ C = \begin{bmatrix} 1 & 0 \end{bmatrix}, \ D = 0. \tag{3.5}$$

Since $D = 0$ holds for many physical systems, we usually omit D when it is zero.

[Example 3.2]   Consider an RLC circuit in **Figure 3.2**, where R [Ω] is a resistor, L [H] is an inductor, C [F] is a capacitor, $v_i(t)$ is the input voltage, $v_R(t)$ is the voltage of the resistor, $v_L(t)$ is the voltage of the inductor, $v_C(t)$ is the voltage of the capacitor, and $i(t)$ is the circuit current. To find the relationship between $v_i(t)$ and $v_C(t)$, first, we recall the relationships between the current and the voltages of the three elements in the circuit

$$i(t) = C\frac{\mathrm{d}v_C(t)}{\mathrm{d}t}, \ v_L(t) = L\frac{\mathrm{d}i(t)}{\mathrm{d}t}, \ v_R(t) = Ri(t). \tag{3.6}$$

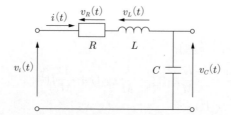

Figure 3.2   RLC circuit

Then, we use Kirchhoff's second law for the circuit

$$v_i(t) = v_L(t) + v_R(t) + v_C(t). \tag{3.7}$$

Substituting (3.6) into (3.7) yields

$$v_i(t) = L\frac{di(t)}{dt} + Ri(t) + v_C(t). \tag{3.8}$$

Using the first relationship in (3.6) eliminates the intermediate variable, $i(t)$, we have

$$v_i(t) = LC\frac{d^2 v_C(t)}{dt^2} + RC\frac{dv_C(t)}{dt} + v_C(t). \tag{3.9}$$

Rearranging terms in (3.9) gives

$$LC\frac{d^2 v_C(t)}{dt^2} + RC\frac{dv_C(t)}{dt} + v_C(t) = v_i(t). \tag{3.10}$$

Choosing $x(t) = [v_C(t),\ \dot{v}_C(t)]^{\mathrm{T}}$ allows us to write the model in the form of (3.1), where

$$A = \begin{bmatrix} 0 & 1 \\ -\dfrac{1}{LC} & -\dfrac{R}{L} \end{bmatrix}, \ B = \begin{bmatrix} 0 \\ \dfrac{1}{LC} \end{bmatrix}, \ C = \begin{bmatrix} 1 & 0 \end{bmatrix}. \tag{3.11}$$

[Example 3.3] Consider the arm robot on a horizontal plane (Figure 1.3). Parameters and variables are as follows (**Figure 3.3**):

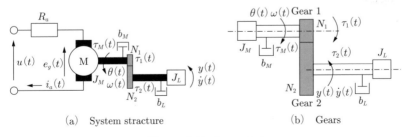

(a) System structure   (b) Gears

Figure 3.3 Arm robot

b_M (b_L) [Nm · s/rad] is the friction coefficient of the motor (load), K_E [Vs/rad] is the back-electromotive-force constant, K_T [Nm/A] is the torque constant, J_M (J_L) [kg · m^2] is the moment of inertia of the motor (load), N_1 (N_2) is the number of teeth of Gear 1 (Gear 2), η is the gear ratio, R_a [Ω] is the resistance of the armature circuit, $e_g(t)$ [V] is the back electromotive force, $i_a(t)$ [A] is the armature current, $u(t)$ [V] is the applied voltage, $y(t)$ [$\theta(t)$] [rad/s] is the rotational angle of the load (motor) shaft, $\dot{y}(t)$ [$\omega(t)$] [rad/s] is the rotational speed of the load (motor) shaft, $\tau_M(t)$ [Nm] is the torque produced on the motor shaft, $\tau_1(t)$ [$\tau_2(t)$] [Nm] is the torque from Gears 1 to 2 (from Gear 2 to the load), and $\theta(t)$ [$y(t)$] [rad] is the rotational angle of the motor shaft (load side). Note that K_T and K_E have the same value for the *SI system of units* (国際単位系).

A mathematical model of the arm robot is given by

Motion:
$$\begin{cases} \text{Load side: } J_L \dfrac{\mathrm{d}^2 y(t)}{\mathrm{d}t^2} = \tau_2(t) - b_L \dfrac{\mathrm{d}y(t)}{\mathrm{d}t}, \\ \text{Motor side: } J_M \dfrac{\mathrm{d}^2 \theta(t)}{\mathrm{d}t^2} = \tau_M(t) - \tau_1(t) - b_M \dfrac{\mathrm{d}\theta(t)}{\mathrm{d}t}, \end{cases} \quad (3.12)$$

Armature circuit: $u(t) = R_a i_a(t) + e_g(t)$, (3.13)

Fleming's left-hand rule: $\tau_M(t) = K_T i_a(t)$, (3.14)

Fleming's right-hand rule: $e_g(t) = K_E \omega(t)$, (3.15)

Gearbox: $\dfrac{\tau_1(t)}{\tau_2(t)} = \dfrac{y(t)}{\theta(t)} = \dfrac{\dot{y}(t)}{\omega(t)} = \dfrac{N_1}{N_2} = \dfrac{1}{\eta}$. (3.16)

Letting

$$J = J_L + \eta^2 J_M, \ b = b_L + \eta^2 b_M \qquad (3.17)$$

and combining equations in (3.12) yield

$$J \dfrac{\mathrm{d}^2 y(t)}{\mathrm{d}t^2} = \eta \tau_M(t) - b \dfrac{\mathrm{d}y(t)}{\mathrm{d}t}. \qquad (3.18)$$

Substituting (3.13) into (3.14) and (3.16) into (3.15) yield

$$\tau_M(t) = K_T \frac{u(t) - e_g(t)}{R_a}, \quad e_g(t) = K_E \omega(t) = \eta K_E \frac{dy(t)}{dt},$$
(3.19)

and substituting (3.16) and (3.19) in to (3.18) yields

$$J\frac{d^2 y(t)}{dt^2} = -\left(b + \frac{\eta^2 K_T K_E}{R_a}\right)\frac{dy(t)}{dt} + \frac{\eta K_T}{R_a}u(t).$$
(3.20)

Choosing a state to be $x(t) = [x_1(t), \ x_2(t)]^{\mathrm{T}} = [y(t), \ \dot{y}(t)]^{\mathrm{T}}$, we have

$$\begin{cases} \dfrac{dx_1(t)}{dt} = x_2(t), \\[2mm] \dfrac{dx_2(t)}{dt} = -\dfrac{1}{J}\left(b + \dfrac{\eta^2 K_T K_E}{R_a}\right)x_2(t) + \dfrac{\eta K_T}{JR_a}u(t), \\[2mm] y(t) = x_1(t). \end{cases}$$
(3.21)

It gives

$$A = \begin{bmatrix} 0 & 1 \\ 0 & -\dfrac{1}{J}\left(b + \dfrac{\eta^2 K_T K_E}{R_a}\right) \end{bmatrix}, \ B = \begin{bmatrix} 0 \\ \dfrac{\eta K_T}{JR_a} \end{bmatrix}, \ C = \begin{bmatrix} 1 & 0 \end{bmatrix}.$$
(3.22)

The state-space model has the following advantages:

i) It describes not only the input-output characteristics of a system but also its internal structure (that is, the relationships between the input, the state, and the output).

ii) Linear algebra is used as a tool to analyze and design a control system.

iii) Both single-input single-output (SISO) and multi-input multi-output (MIMO) systems can be handled in the same way.

3.2 Transfer Function

A spring, a mass, and a damper are basic mechanical elements for rectilinear motions (**Figure 3.4**), where K [N/m] is an elastic modulus, M [kg] is a mass, b [N · s/m] is a viscous-friction coefficient, $f(t)$ [N] is a force, and $y(t)$ [m] is a displacement. Applying Newton's second law of motion yields the *equations of motion* (運動方程式) of those elements (Figure 3.4), which have been shown in Example 3.1.

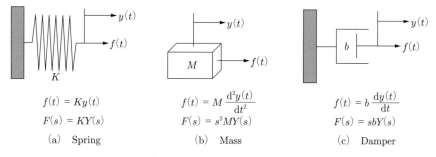

$$f(t) = Ky(t)$$
$$F(s) = KY(s)$$
(a) Spring

$$f(t) = M\frac{\mathrm{d}^2y(t)}{\mathrm{d}t^2}$$
$$F(s) = s^2MY(s)$$
(b) Mass

$$f(t) = b\frac{\mathrm{d}y(t)}{\mathrm{d}t}$$
$$F(s) = sbY(s)$$
(c) Damper

Figure 3.4 Equations of rectilinear motions of basic mechanical elements

Classical control theory uses a transfer function to analyze and synthesize a linear system. Taking the Laplace transform for each term in Figure 3.4 and using the property of differentiation for zero initial values[†] [that is, $\mathcal{L}\{f^{(n)}(t)\} = s^nF(s)$ for $f(0) = \dot{f}(0) = f^{(n-1)}(0) = 0$] yield the input-output relationships in the s domain for those elements, which are also shown in the same figure.

The relationships in the time and the s domains between a torque $\tau(t)$ [Nm] and a rotational angle $\theta(t)$ [rad] are shown in **Figure 3.5** for the common mechanical elements for rotational motions, where K_θ [Nm/rad] is an elastic modulus, J [kg · m^2] is the moment of inertia, and C_θ [Nm · s/rad]

[†]   The initial values of variables are usually set to be zero in classical control theory.

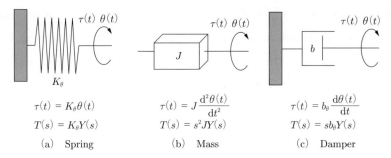

$$\tau(t) = K_\theta \theta(t)$$
$$T(s) = K_\theta Y(s)$$

(a) Spring

$$\tau(t) = J\frac{d^2\theta(t)}{dt^2}$$
$$T(s) = s^2 JY(s)$$

(b) Mass

$$\tau(t) = b_\theta \frac{d\theta(t)}{dt}$$
$$T(s) = sb_\theta Y(s)$$

(c) Damper

Figure 3.5 Equations of rotational motions of basic mechanical elements

is a viscous-friction coefficient.

The relationships between voltages and currents for common electric elements in the time and the s domains are shown in **Figure 3.6**, where R [Ω] is a resistor; L [H] is an inductor; C [F] is a capacitor; $i(t)$ [A] is the current that flows through an electric element; and $v_R(t)$ [V], $v_L(t)$ [V], and $v_C(t)$ [V] are the voltages of the resistor, capacitor, and inductor, respectively.

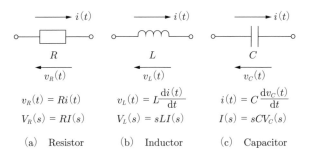

$$v_R(t) = Ri(t)$$
$$V_R(s) = RI(s)$$

(a) Resistor

$$v_L(t) = L\frac{di(t)}{dt}$$
$$V_L(s) = sLI(s)$$

(b) Inductor

$$i(t) = C\frac{dv_C(t)}{dt}$$
$$I(s) = sCV_C(s)$$

(c) Capacitor

Figure 3.6 Mathematical models of common electric elements

Mechanical-electrical analogies　Since a control system is designed based on a mathematical model of a plant. If the model of an electric system is exactly the same as that of a mechanic system, we can treat them in the same way and use the same method to design a control system. This way of thinking not only is important to establish a theoretical system for control engineering but also helps us explain phenomena of one unfamiliar field in one of our familiar fields. Let $q(t)$ be an electric charge and C_m be the compliance $(= 1/K)$. The mechanical-electrical analogies are shown in **Table 3.1**.

Table 3.1　Mechanical-electrical analogies

Electricity	$v(t) = \dfrac{1}{C}q(t)$	$v(t) = R\dfrac{\mathrm{d}q(t)}{\mathrm{d}t}$	$v(t) = L\dfrac{\mathrm{d}^2q(t)}{\mathrm{d}t^2}$
Machinery	$f(t) = \dfrac{1}{C_m}y(t)$	$f(t) = b\dfrac{\mathrm{d}y(t)}{\mathrm{d}t}$	$f(t) = M\dfrac{\mathrm{d}^2y(t)}{\mathrm{d}t^2}$

An electric current is $i(t) = \mathrm{d}q(t)/\mathrm{d}t$ and a speed is $v(t) = \mathrm{d}y(t)/\mathrm{d}t$. Let an impedance be $Z(s) = V(s)/I(s)$ for an electric system and be $Z(s) = F(s)/V(s)$ for a mechanic system[†]. Impedance analogies for electric elements [input: $i(t)$, output: $v(t)$] and those for mechanic elements [input: $v(t)$, output: $f(t)$] are shown in **Table 3.2**. An admittance is defined to be the inverse of an impedance.

Table 3.2　Impedance analogies

Electricity	$Z(s) = \dfrac{V(s)}{I(s)}$	Machinery	$Z(s) = \dfrac{F(s)}{V(s)}$
$\dfrac{\mathrm{d}v(t)}{\mathrm{d}t} = \dfrac{1}{C}i(t)$	$\dfrac{1}{sC}$	$\dfrac{\mathrm{d}f(t)}{\mathrm{d}t} = \dfrac{1}{C_m}v(t)$	$\dfrac{1}{sC_m}$
$v(t) = Ri(t)$	R	$f(t) = bv(t)$	b
$v(t) = L\dfrac{\mathrm{d}i(t)}{\mathrm{d}t}$	sL	$f(t) = M\dfrac{\mathrm{d}v(t)}{\mathrm{d}t}$	sM

It is called impedance control in mechanical engineering, robotics, and biomechanics if the input of a system is speed (or displacement) and the output is force; and is called admittance control vice versa.

[Example 3.4]　Consider the spring-mass-damper system (Figure 3.1). Carrying out the Laplace transform for (3.2) for zero initial conditions

---

[†]　The same variable $v(t)$ is used as a voltage in an electric system and a speed in a mechanical system. Since its meaning is clear from the context, this should not cause confusion.

yields

$$Ms^2Y(s) = F(s) - KY(s) - bsY(s) \qquad (3.23)$$

where $F(s) = \mathcal{L}\{f(t)\}$, $Y(s) = \mathcal{L}\{y(t)\}$, and $\dot{y}(0) = y(0) = 0$. Thus,

$$(Ms^2 + bs + K)Y(s) = F(s).$$

Note that the division between $Y(s)$ and $F(s)$, $G(s) = Y(s)/F(s)$, depends on neither the input nor[†1] the output signals. It contains all the parameters of the system. Thus, it can be used to describe system characteristics. $G(s)$ is called a *transfer function* (伝達関数). We obtain the transfer function of the system (3.2) as

$$G(s) = \frac{Y(s)}{F(s)} = \frac{1}{Ms^2 + bs + K}. \qquad (3.24)$$

【Example 3.5】 Consider the RLC circuit (Figure 3.2). Performing the Laplace transform for (3.10) yields

$$LCs^2V_C(s) + RCsV_C(s) + V_C(s) = V_i(s), \qquad (3.25)$$

where $V_i(s) = \mathcal{L}\{v_i(t)\}$ and $V_C(s) = \mathcal{L}\{v_C(t)\}$. It follows that[†2] the transfer function from $v_i(t)$ to $v_C(t)$ is

$$G(s) = \frac{V_C(s)}{V_i(s)} = \frac{1}{LCs^2 + RCs + 1}. \qquad (3.26)$$

【Example 3.6】 Consider the arm robot on a horizontal plane (Figure 3.3). Performing the Laplace transform for (3.20) yields the transfer function of the arm robot

---

†1   neither A nor B：A も B も～ない。
†2   it follows that ~：[当然の結果として]~ということになる。

$$P(s) = \frac{\beta}{s(s+\alpha)}, \quad \alpha = \frac{1}{J}\left(b + \frac{\eta^2 K_T K_E}{R_a}\right), \quad \beta = \frac{\eta K_T}{J R_a}. \quad (3.27)$$

Generally speaking, a model of a *linear system* (線形システム) is usually described by an *ordinary differential equation* (常微分方程式), which is given by

$$\frac{d^n y(t)}{dt^n} + a_{n-1}\frac{d^{n-1}y(t)}{dt^{n-1}} + \cdots + a_1\frac{dy(t)}{dt} + a_0 y(t)$$

$$= b_m\frac{d^m u(t)}{dt^m} + b_{m-1}\frac{d^{m-1}u(t)}{dt^{m-1}} + \cdots + b_1\frac{du(t)}{dt} + b_0 u(t), \quad (3.28)$$

where $m \leqq n$; $u(t)$ and $y(t)$ are the input and output of the system, respectively; and a_i $(i = 0, 1, \cdots, n)$ and b_j $(j = 0, 1, \cdots, m)$ are constants.

Taking the Laplace transform for each term in (3.28) for zero initial values yields

$$s^n Y(s) + a_{n-1}s^{n-1}Y(s) + \cdots + a_1 s Y(s) + a_0 Y(s)$$

$$= b_m s^m U(s) + b_{m-1}s^{m-1}U(s) + \cdots + b_1 s U(s) + b_0 U(s). \quad (3.29)$$

The transfer function of the system (3.28) is

$$G(s) = \frac{Y(s)}{U(s)} = \frac{b_m s^m + b_{m-1}s^{m-1} + \cdots + b_1 s + b_0}{s^n + a_{n-1}s^{n-1} + \cdots + a_1 s + a_0}. \quad (3.30)$$

If we define two polynomial functions of s, $N(s)$ and $D(s)$, then $G(s)$ is a rational function of those two

$$\begin{cases} G(s) = \dfrac{N(s)}{D(s)}, \\ N(s) = b_m s^m + b_{m-1}s^{m-1} + \cdots + b_1 s + b_0, \\ D(s) = s^n + a_{n-1}s^{n-1} + \cdots + a_1 s + a_0. \end{cases} \quad (3.31)$$

The roots of $N(s) = 0$ is called the *zeros* (零点, ゼロ点) of the system; and those of $D(s) = 0$, the *poles* (極).

The output of the system is simply given by $Y(s) = G(s)U(s)$. Clear, this description is much simpler than that given by (3.28). The features of using a transfer function are as follows:

i) Since $G(s)$ is a rational function of the complex variable s, we can use complex analysis to analyze the stability and the characteristics of a system.

ii) Not only the steady-state response but also the transient behavior of a system can be analyzed based on the poles and zeros of a system.

iii) Many tools were devised to analyze and synthesize a system graphically or computationally.

3.3 Linear Approximation Model

We need to derive a linear model for a nonlinear system to use linear control theory. Note that many control systems operate around operating points, an easy way to build a linear model is to carry out linearization at an operating point.

The Taylor series of a nonlinear function $y = f(x)$ at x_0 is

$$f(x) = f(x_0) + \dot{f}(x_0)(x - x_0) + \frac{\ddot{f}(x_0)}{2!}(x - x_0)^2$$
$$+ \cdots + \frac{f^{(n)}(x_0)}{n!}(x - x_0)^n + \cdots . \qquad (3.32)$$

Carrying out a linear approximation of the function at $x = x_0$ yields

$$y = f(x_0) + \dot{f}(x_0)(x - x_0). \qquad (3.33)$$

Letting $\Delta x = x - x_0$, $\Delta y = f(x) - f(x_0)$, and $K = \dot{f}(x_0)$, we have

$$\Delta y = K \Delta x. \qquad (3.34)$$

Clearly, it is a linear function[†].

[†] Note that a function is linear if it satisfies additivity $[f(x_a + x_b) = f(x_a) + f(x_b)]$ and homogeneity $[f(kx) = kf(x)]$. So, $y = Kx + b$ is a linear function of x iff $b = 0$.

A multivariate version of the Taylor series is derived for a multivariate function $y = f(x_1, x_2, \cdots, x_n)$ at $(x_{10}, x_{20}, \cdots, x_{n0})$. Let $x = [x_1, x_2, \cdots, x_n]^{\mathrm{T}}$. The linear approximation of the function at $x_0 = [x_{10}, x_{20}, \cdots, x_{n0}]^{\mathrm{T}}$ is

$$f(x) = f(x_0) + \left.\frac{\partial f(x)}{\partial x_1}\right|_{x=x_0} (x_1 - x_{10}) + \left.\frac{\partial f(x)}{\partial x_2}\right|_{x=x_0} (x_2 - x_{20})$$
$$+ \cdots + \left.\frac{\partial f(x)}{\partial x_n}\right|_{x=x_0} (x_n - x_{n0}). \tag{3.35}$$

Now, we consider a nonlinear system

$$\begin{cases} \dfrac{\mathrm{d}x(t)}{\mathrm{d}t} = f(x(t), u(t)), \\ y(t) = h(x(t)), \end{cases} \tag{3.36}$$

where $x(t)$ is the state, $u(t)$ is the input, and $y(t)$ is the output. Let (x_0, u_0) be an equilibrium, that is,

$$f(x_0, u_0) = 0. \tag{3.37}$$

The linear approximation of the system (3.36) at $(x_0,\ u_0)$ is

$$\begin{cases} f(x, u) = f(x_0, u_0) + A(x - x_0) + B(u - u_0), \\ h(x) = h(x_0) + C(x - x_0), \\ A = \left.\dfrac{\partial f(x, u)}{\partial x}\right|_{\substack{x=x_0 \\ u=u_0}}, \quad B = \left.\dfrac{\partial f(x, u)}{\partial u}\right|_{\substack{x=x_0 \\ u=u_0}}, \quad C = \left.\dfrac{\mathrm{d}h(x)}{\mathrm{d}x}\right|_{x=x_0}. \end{cases} \tag{3.38}$$

Letting

$$\Delta x(t) = x(t) - x_0, \ \Delta u(t) = u(t) - u_0, \ \Delta y(t) = y(t) - h(x_0) \tag{3.39}$$

and substituting (3.37)~(3.39) into (3.36) yield a linear approximation model of the system (3.36)

$$\begin{cases} \dfrac{\mathrm{d}\Delta x(t)}{\mathrm{d}t} = A\Delta x(t) + B\Delta u(t), \\ \Delta y(t) = C\Delta x(t). \end{cases} \tag{3.40}$$

[Example 3.7] Consider an arm robot on a vertical plane (**Figure 3.7**), in which J [kg · m^2] is the moment of inertia of the arm, l [m] is the distance from the center of gravity to the center of rotation, b [Nm · s/rad] is the friction coefficient of the arm, $\tau(t)$ [Nm] is the torque added to the arm, and $\theta(t)$ [rad] is the rotational angle of the arm.

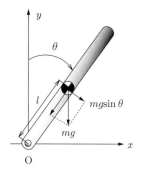

Figure 3.7 Arm robot on vertical plane

The control objective is to maintain the rotational angle at a prescribed value θ_0.

We need to establish a mathematical model to carry out control. The equation of motion is

$$J\frac{\mathrm{d}^2\theta(t)}{\mathrm{d}t^2} = \tau(t) - b\frac{\mathrm{d}\theta(t)}{\mathrm{d}t} - mgl\sin\theta(t). \qquad (3.41)$$

It is nonlinear because it contains a sine function of the rotational angle. We derive a linear approximation model of (3.41) so as to[†] use the linear control theory to design a controller. Let

$$\theta(t) = \theta_0 + \Delta\theta(t), \ \tau(t) = \tau_0 + \Delta\tau(t) \qquad (3.42)$$

and note that

$$\left.\frac{\mathrm{d}^2\theta(t)}{\mathrm{d}t^2}\right|_{\theta(t)=\theta_0} = \left.\frac{\mathrm{d}\theta(t)}{\mathrm{d}t}\right|_{\theta(t)=\theta_0} = 0, \ \tau_0 = mgl\sin\theta_0. \qquad (3.43)$$

Since (3.41) only has one nonlinearity, we carry out linear approximation for it using the Taylor series

$$\sin(\theta_0 + \Delta\theta(t)) = \sin\theta_0 + \cos\theta_0 \cdot \Delta\theta(t). \qquad (3.44)$$

Combining (3.41)~(3.44) yields a linear approximation model

---

† so as to ~ : ~するように，~するために。

$$J\frac{\mathrm{d}^2\Delta\theta(t)}{\mathrm{d}t^2} = \Delta\tau(t) - b\frac{\mathrm{d}\Delta\theta(t)}{\mathrm{d}t} - mgl\cos\theta_0 \cdot \Delta\theta(t). \quad (3.45)$$

Letting $x(t) = [\Delta\theta(t), \mathrm{d}\Delta\theta(t)/\mathrm{d}t]^{\mathrm{T}}$, $u(t) = \Delta\tau(t)$, and $y(t) = \Delta\theta(t)$ gives the state-space representation of the linear model

$$A = \begin{bmatrix} 0 & 1 \\ -\dfrac{mgl\cos\theta_0}{J} & -\dfrac{b}{J} \end{bmatrix}, \ B = \begin{bmatrix} 0 \\ \dfrac{1}{J} \end{bmatrix}, \ C = \begin{bmatrix} 1 & 0 \end{bmatrix}. \quad (3.46)$$

Its transfer function is

$$P(s) = \frac{\Delta\Theta(s)}{\Delta T(s)} = \frac{1}{Js^2 + bs + mgl\cos\theta_0}. \quad (3.47)$$

We often use the following approximations

$$\sin\theta \doteqdot \theta, \quad \cos\theta \doteqdot 1 \quad (3.48)$$

for a small θ in control engineering. However, as shown in **Table 3.3**, (3.48) are good approximations even for angles that are not very small.

Table 3.3 Sine and cosine functions

θ [deg]	θ [rad]	$\sin\theta$	$\cos\theta$
1	0.01745	0.01745	0.9998
5	0.08726	0.08716	0.9962
10	0.1745	0.1736	0.9848
15	0.2618	0.2588	0.9659
20	0.3490	0.3420	0.9397
25	0.4363	0.4226	0.9063
30	0.5236	0.5000	0.8660
35	0.6109	0.5736	0.8192
40	0.6981	0.6428	0.7660
45	0.7854	0.7071	0.7071

[Example 3.8] We use an inverted pendulum on a cart as an example to show how to establish a linear model for a nonlinear system.

Let m_c [kg] be the mass of the cart; m_p [kg], the mass of the pendulum; J_p [kg $\cdot$ m^2], the moment of inertia of the pendulum; b_c [kg/s], the viscous-friction coefficient of the cart; b_θ [kg $\cdot$ m^2/s], the viscous friction coefficient of the pendulum; and l_p [m], the distance from the pivot to the center of gravity of the pendulum.

Let $z(t)$ be the cart position from a reference point and $\theta(t)$ be the pendulum angle to the vertical axis (**Figure 3.8**). The equation of the linear motion of the cart is

$$m_c \ddot{z}(t) = f_c(t) - b_c \dot{z}(t) - H(t). \tag{3.49}$$

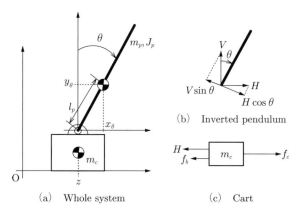

(a)   Whole system   (c)   Cart

Figure 3.8   Inverted pendulum on cart

The equations of the linear motion of the pendulum are

$$\begin{cases} m_p \ddot{x}_\theta(t) = H(t), \\ m_p \ddot{y}_\theta(t) = V(t) - m_p g, \end{cases} \tag{3.50}$$

and the equation of the rotational motion of the pendulum around its center of gravity is

$$J_p \ddot{\theta}(t) = -b_\theta \dot{\theta}(t) + V(t) l_p \sin\theta(t) - H(t) l_p \cos\theta(t). \tag{3.51}$$

In (3.49)~(3.51), $f_c(t)$ [N] is an input force; $H(t)$, the horizontal action and reaction force between the cart and the pendulum; and $V(t)$, the vertical one. Note that

$$\begin{cases} x_\theta(t) = z(t) + l_p \sin \theta(t), \\ y_\theta(t) = l_p \cos \theta(t). \end{cases} \tag{3.52}$$

Eliminating $H(t)$ and $V(t)$ in (3.49)~(3.51) gives the model of the system

$$\begin{cases} (m_c + m_p)\ddot{z}(t) + m_p l_p \cos \theta(t) \cdot \ddot{\theta}(t) \\ \quad = f_c(t) - b_c \dot{z}(t) + m_p l_p \dot{\theta}^2(t) \sin \theta(t), \\ m_p l_p \cos \theta(t) \cdot \ddot{z}(t) + (J_p + m_p l_p^2)\ddot{\theta}(t) \\ \quad = -b_\theta \dot{\theta}(t) + m_p g l_p \sin \theta(t). \end{cases} \tag{3.53}$$

Since the model contains the nonlinear items of $\cos \theta(t) \cdot \ddot{\theta}(t)$, $\dot{\theta}^2(t) \sin \theta(t)$, $\cos \theta(t) \cdot \ddot{z}(t)$, and $\sin \theta(t)$, it is a nonlinear model. A control problem for this system is to stabilize it at $\theta(t) = 0$. So, we build a linear model at $\theta(t) = 0$ as follows.

The approximations (3.48) and

$$\dot{\theta}^2(t) \sin \theta(t) \doteq 0, \quad \dot{\theta}^2(t)\theta(t) \doteq 0 \tag{3.54}$$

hold around $\theta(t) = 0$ and $\dot{\theta}(t) = 0$. As a result, we obtain a linear model of the system

$$\begin{cases} (m_c + m_p)\ddot{z}(t) + m_p l_p \ddot{\theta}(t) = f_c(t) - b_c \dot{z}(t), \\ m_p l_p \ddot{z}(t) + (J_p + m_p l_p^2)\ddot{\theta}(t) = -b_\theta \dot{\theta}(t) + m_p g l_p \theta(t). \end{cases} \tag{3.55}$$

That is,

$$\begin{bmatrix} m_c + m_p & m_p l_p \\ m_p l_p & J_p + m_p l_p^2 \end{bmatrix} \begin{bmatrix} \ddot{z}(t) \\ \ddot{\theta}(t) \end{bmatrix} = \begin{bmatrix} f_c(t) - b_c \dot{z}(t) \\ -b_\theta \dot{\theta}(t) + m_p g l_p \theta(t) \end{bmatrix}. \tag{3.56}$$

Solving (3.56) for $\ddot{z}(t)$ and $\ddot{\theta}(t)$ gives

$$\begin{bmatrix} \ddot{z}(t) \\ \ddot{\theta}(t) \end{bmatrix} = \begin{bmatrix} \dfrac{\Lambda_1}{\Xi} \\ \dfrac{\Lambda_2}{\Xi} \end{bmatrix}, \tag{3.57}$$

where

$$\begin{cases} \Xi = (m_c + m_p)J_p + m_c m_p l_p^2, \\ \Lambda_1 = (J_p + m_p l_p^2)f_c(t) - (J_p + m_p l_p^2)b_c \dot{z}(t) \\ \quad + m_p l_p b_p \dot{\theta}(t) - m_p^2 l_p^2 g \theta(t), \\ \Lambda_2 = -m_p l_p f_c(t) + m_p l_p b_c \dot{z}(t) - (m_c + m_p)b_\theta \dot{\theta}(t) \\ \quad + (m_c + m_p)m_p l_p g \theta(t). \end{cases}$$

Letting

$$\begin{cases} x(t) = [z(t),\ \theta(t),\ \dot{z}(t),\ \dot{\theta}(t)]^{\mathrm{T}}, \\ u(t) = f_c(t),\ y(t) = [z(t),\ \theta(t)]^{\mathrm{T}}, \end{cases}$$

we have the state-space form of (3.57)

$$\begin{cases} A = \begin{bmatrix} 0 & 0 & 1 & 0 \\ 0 & 0 & 0 & 1 \\ 0 & -\dfrac{m_p^2 l_p^2 g}{\Xi} & -\dfrac{(J_p + m_p l_p^2)b_c}{\Xi} & \dfrac{m_p l_p b_\theta}{\Xi} \\ 0 & \dfrac{(m_c + m_p)m_p l_p g}{\Xi} & \dfrac{m_p l_p b_c}{\Xi} & -\dfrac{(m_c + m_p)b_\theta}{\Xi} \end{bmatrix}, \\[6pt] B = \begin{bmatrix} 0 \\ 0 \\ \dfrac{J_p + m_p l_p^2}{\Xi} \\ -\dfrac{m_p l_p}{\Xi} \end{bmatrix},\ C = \begin{bmatrix} 1 & 0 & 0 & 0 \\ 0 & 1 & 0 & 0 \end{bmatrix}. \end{cases}$$

$$(3.58)$$

Problems

Basic Level

[1] Find the state-space representations of systems in **Figure 3.9**. In each system, the input is a displacement $r(t)$ or an applied force $f(t)$ and the output is a displacement $y(t)$. $x(t)$ in Figure 3.9(b) is the displacement on $f(t)$ side.

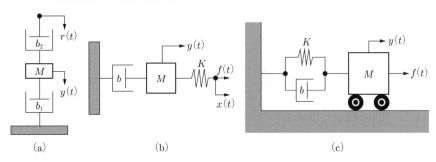

Figure 3.9   Problems 1 and 6

〔2〕   Find the state-space representations of systems in **Figure 3.10**.

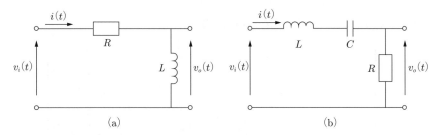

Figure 3.10   Problems 2 and 7

〔3〕   When we calculate a transfer function, we usually assume that the initial condition of the system is zero.

    A.   True   B.   False.

〔4〕   The transfer function of a linear system is the Laplace transform of the output of the system for the input of a

    A.   step signal   B.   ramp signal

    C.   impulse signal   D.   sinusoidal signal.

〔5〕   In a mechanical system, a spring force is proportional to

    A.   velocity   B.   displacement   C.   acceleration   D.   all of them.

〔6〕   Find the differential equations and the transfer functions of systems in Figure 3.9.

〔7〕   Find the differential equations and the transfer functions of systems in Figure 3.10, where the inputs and outputs are voltages $v_i(t)$ and $v_o(t)$, respectively.

[8]   Find the original differential equation of the system $G(s) = Y(s)/U(s) = s/(s+1)(s+2)$.

[9]   Find the mathematical models and the transfer functions of systems in **Figure 3.11** $[G(s) = Y(s)/F(s)]$. $x(t)$ is the displacement on the right side of K_1 [Figure 3.11(a)] or b_1 [Figure 3.11(b)].

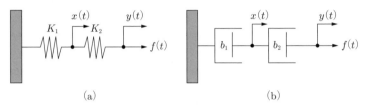

(a)   (b)

Figure 3.11   Problem 9

[10]   Find the transfer function $G(s) = V_o(s)/V_i(s)$ in **Figure 3.12** for $R_1 = 1\,\Omega$, $R_2 = 1\,\Omega$, $L = 1\,\mathrm{H}$, and $C = 1\,\mathrm{F}$.

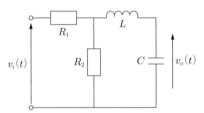

Figure 3.12   Problem 10

[11]   Consider a nozzle-flapper mechanism in an electro-hydraulic servo valve (**Figure 3.13**). Let the supply pressure be $p_s(t)$; the backpressure in the nozzle be $p_1(t)$; the input flow be $q_1(t)$; the output flow be $q_2(t)$; the distance between the nozzle and the flapper be $x(t)$; the fluid density be ρ; the aperture area and the discharge coefficient of the orifice be a and c_1, respectively; and the pore size and the discharge coefficient of the nozzle be d and c_2, respectively. $q_1(t) = q_2(t)$ provides $c_1 a \sqrt{\dfrac{2}{\rho}[p_s(t) - p_1(t)]} = c_2 \pi d x(t) \sqrt{\dfrac{2}{\rho} p_1(t)}$, that is, $p_1(t) = \dfrac{p_s(t)}{1 + wx^2(t)}$, where $w = (c_2 \pi d/c_1 a)^2$. Let $p_s(t)$ be a constant $[p_s(t) = p_{s0}]$, Find a linear approximation model of the relationship between $p_1(t)$ and $x(t)$ at $x(t) = x_0$.

[12]   Consider a liquid-level control problem in **Figure 3.14**. Let the amounts of the input and output flows be $q_i(t)$ and $q_o(t)$, respectively; the height of

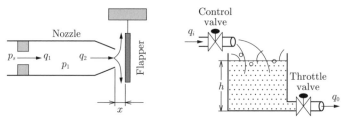

Figure 3.13 Problem 11   **Figure 3.14** Problem 12

the tank be $h(t)$; the cross-section area of the tank be S; and the discharge
coefficient of the throttle valve be c. The law of conservation of mass gives
$Sdh(t) = [q_i(t) - q_o(t)]dt$ or $\dfrac{dh(t)}{dt} = \dfrac{q_i(t) - q_o(t)}{S}$. The characteristic of
the throttle valve gives $q_o(t) = c\sqrt{h(t)}$. Thus, the model for liquid-level
control is $\dfrac{dh(t)}{dt} + \dfrac{c}{S}\sqrt{h(t)} = \dfrac{1}{S}q_i(t)$. Find its linear approximation model
at $h(t) = h_0$ and present it as a state-space model and a transfer function.

Advanced Level

【1】 Mathematical modeling of control systems may be achieved by the use of
   A. differential equations   B. transfer functions
   C. state-space representation   D. all of the above.

【2】 Find the state-space representations of systems in **Figure 3.15**. The input
   and output are applied force $f(t)$ and displacement of Cart 2 $[x_2(t)]$ in
   Figure 3.15(a), and applied force $f(t)$ and the displacement of Cart 1 $[x_1(t)]$
   in Figure 3.15(b), respectively.

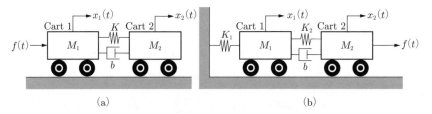

(a)   (b)

Figure 3.15 Advanced problem 2

【3】 Find the transfer function of a system in **Figure 3.16** where the input is
   $v_i(t)$ and the output is $v_o(t)$.

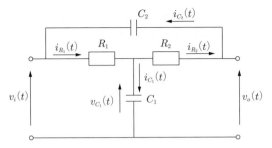

Figure 3.16   Advanced problem 3

【4】 A TORA (translational oscillator with a rotational actuator) system (**Figure 3.17**) is a simplified model of a dual-spin spacecraft. It consists of a cart and an eccentric rotational proof mass. The cart moves horizontally along a line and is connected to a wall by a spring. The proof mass is attached to the cart and is actuated by a DC motor. $q_1(t)$ is the translational position of the cart, $q_2(t)$ is the rotational angle of the proof mass, m_1 is the mass of the cart, m_2 is the mass of the proof mass, r is the rotary radius of the proof mass, J is the moment of inertia of the proof mass, k is a spring constant, and $\tau_2(t)$ is the control torque that drives the proof mass. Verify that the dynamics of a TORA is

$$M(q_2(t))\begin{bmatrix}\ddot{q}_1(t)\\\ddot{q}_2(t)\end{bmatrix}+\begin{bmatrix}-m_2r\dot{q}_2^2(t)\sin q_2(t)\\0\end{bmatrix}+\begin{bmatrix}kq_1(t)\\0\end{bmatrix}=\begin{bmatrix}0\\\tau_2(t)\end{bmatrix},$$
(3.59)

where

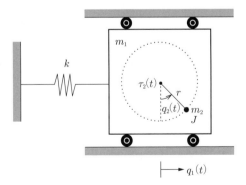

Figure 3.17   TORA

$$M(q_2(t)) = \begin{bmatrix} m_1 + m_2 & m_2 r \cos q_2(t) \\ m_2 r \cos q_2(t) & m_2 r^2 + J \end{bmatrix}.$$

Consider the TORA system

$$\begin{cases} m_1 = 4.9 \, \text{kg}, & m_2 = 0.1 \, \text{kg}, & r = 0.8 \, \text{m} \\ J = 0.064 \, \text{kg} \cdot \text{m}^2, & k = 350 \, \text{N/m} \end{cases}$$

Let the output be $y(t) = [q_1(t), \; q_2(t)]^{\mathrm{T}}$. Find a linear approximation state-space model of (3.59) for $x(t) = [q_1(t), \; q_2(t), \; \dot{q}_1(t), \; \dot{q}_2(t)]^{\mathrm{T}}$ at the origin.

[5] A two-link arm robot moves in a vertical plane (**Figure 3.18**). For the j-th link $(j = 1, 2)$, m_j is the mass, L_j is the length, L_{gj} is the distance from the j-th joint to the center of mass of the link, I_j is the moment of inertia of the link about its centroid, $q_j(t)$ is the angle, $\dot{q}_j(t)$ is the angular velocity, and $\tau_j(t)$ is the torque applied to the j-th joint. g is the gravitational acceleration.

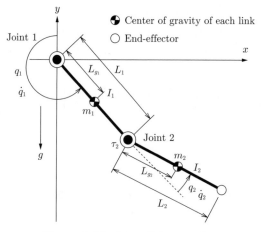

Figure 3.18 A two-link arm robot

Verify that the dynamics of the robot is

$$M(q(t))\ddot{q}(t) + H(q(t), \dot{q}(t)) + G(q(t)) = \tau(t) \tag{3.60}$$

where $\tau(t) = [\tau_1(t) \; \tau_2(t)]^{\mathrm{T}}$ are the applied torques, $M(q(t))$ is the inertia matrix, $H(q(t), \dot{q}(t))$ is the combination of the Coriolis and centrifugal forces, and $G(q(t))$ is the force of gravity. They are

$$
\begin{cases}
m_{11}(q(t)) = m_1 L_{g1}^2 + I_1 + m_2 L_1^2 + m_2 L_{g2}^2 + I_2 + 2 m_2 L_1 L_{g2} \cos q_2(t), \\[4pt]
m_{12}(q(t)) = m_2 L_{g2}^2 + I_2 + m_2 L_1 L_{g2} \cos q_2(t), \\[4pt]
g_1(q(t)) = -(m_1 L_{g1} + m_2 L_1) g \sin q_1(t) - m_2 g L_{g2} g \sin(q_1(t) + q_2(t)), \\[4pt]
M(q(t)) = \begin{bmatrix} m_{11}(q(t)) & m_{12}(q(t)) \\ m_{12}(q(t)) & m_2 L_{g2}^2 + I_2 \end{bmatrix}, \\[16pt]
H(q(t), \dot{q}(t)) = \begin{bmatrix} -m_2 L_1 L_{g2} \left[2\dot{q}_1(t)\dot{q}_2(t) + \dot{q}_2^2(t) \right] \sin q_2(t) \\ m_2 L_1 L_{g2} \dot{q}_1^2(t) \sin q_2(t) \end{bmatrix}, \\[16pt]
G(q(t)) = \begin{bmatrix} g_1(q(t)) \\ -m_2 L_{g2} g \sin(q_1(t) + q_2(t)) \end{bmatrix}.
\end{cases}
$$

$$(3.61)$$

Find a linear approximation state-space model of (3.60) at $x(t) = [q_1(t),\ q_2(t),\ \dot{q}_1(t),\ \dot{q}_2(t)]^{\mathrm{T}} = [0,\ 0,\ 0,\ 0]^{\mathrm{T}}$.

4 | Block Diagram and Frequency Response

As explained in Chapter 3, the input-output relationship of a system can be described by a differential equation in the time domain and can also be described by a transfer function in the s domain. A transfer function has many features. For example, we can analyze a system based on the definitions of poles and zeros that are defined based on the roots of polynomials in s. Moreover, if we use a block to describe the transfer function of an element, then it is easy to use a block diagram to describe an entire system.

Fourier analysis shows that a time-domain signal can be represented by the sum of sine waves. Investigating the characteristics of a system to a sinusoidal input with different angular frequencies is called frequency-response analysis. The response is closely related to the poles and zeros of a system. Understanding the frequency response of a transfer function lays the foundation for classical control theory.

4.1   Equivalent Conversion of Block Diagram

A block element in a block diagram is described as a box with two lines. The transfer function of the element is placed in the box, and an arrow in each of the lines connected to the box shows the directions of signals to and from the element (the input and output, respectively). A white circle is a *summing point* (加算点), where signals are added together. A black circle is a *pick-off point* (抽出点), where the signal is extracted (**Figure 4.1**).

Three kinds of elementary block connections are a *series connection* (直

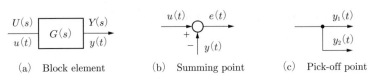

(a) Block element   (b) Summing point   (c) Pick-off point

Figure 4.1 Basic elements of block diagram

列結合), a *parallel connection* (並列結合), and a *feedback connection* (フィードバック結合) **(Figure 4.2)**.

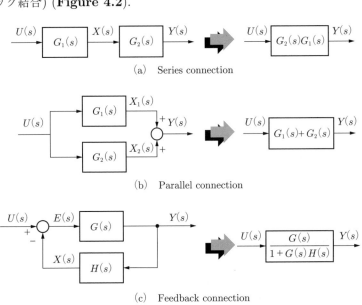

(a) Series connection

(b) Parallel connection

(c) Feedback connection

Figure 4.2 Three kinds of elementary block connections

(1) Series connection   The output is

$$Y(s) = G_2(s)X(s) = G_2(s)G_1(s)U(s). \tag{4.1}$$

Therefore, the transfer function is

$$G(s) = G_2(s)G_1(s). \tag{4.2}$$

(2) Parallel connection   The output is

$$Y(s) = X_1(s) + X_2(s) = G_1(s)U(s) + G_2(s)U(s)$$
$$= [G_1(s) + G_2(s)]U(s).$$

Therefore, the transfer function is

$$G(s) = G_1(s) + G_2(s). \qquad (4.3)$$

(3) Feedback connection   The relationships between the signals are

$$Y(s) = G(s)E(s), \ \ E(s) = R(s) - X(s), \ \ X(s) = H(s)Y(s).$$

Combining the above equations gives

$$Y(s) = G(s)[R(s) - H(s)Y(s)].$$

Solving it for $Y(s)$ yields

$$G_{FB}(s) = \frac{Y(s)}{R(s)} = \frac{G(s)}{1 + G(s)H(s)}. \qquad (4.4)$$

Other basic block-diagram operations are shown in **Figures 4.3** and **4.4**.

(4) Pick-off point (Figure 4.3)

$$Y_2(s) = U(s)$$

equals

$$Y_2(s) = \frac{1}{G(s)}G(s)U(s) = \frac{1}{G(s)}Y(s). \qquad (4.5)$$

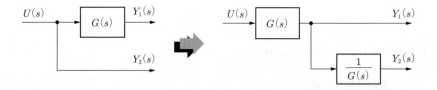

(a)   Before moving   (b)   After moving

Figure 4.3 Moving pick-off point

(5) Summing point (Figure 4.4)

$$Y(s) = G(s)X(s) = G(s)[U(s) + V(s)] = G(s)U(s) + G(s)V(s)$$

equals

$$Y(s) = Y_1(s) + Y_2(s) = G(s)U(s) + G(s)V(s). \tag{4.6}$$

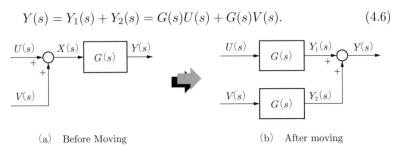

(a) Before Moving   (b) After moving

Figure 4.4 Moving summing point

【Example 4.1】 Draw the block diagram of the arm robot in Figure 3.3. Combining (3.13)~(3.16) and (3.18) yields the block diagram of the arm robot in **Figure 4.5**.

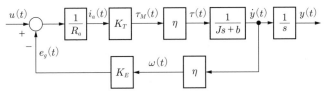

Figure 4.5 Block diagram of arm robot

4.2 Frequency Response

Since we can consider an input signal to be the sum of sine waves, examining the response of a system to a sinusoidal input, which is called the *frequency response* (周波数応答) of the system, is important. Consider a system

$$G(s) = \frac{b_0 s^m + b_1 s^{m-1} + \cdots + b_{m-1}s + b_m}{(s - p_1)(s - p_2) \cdots (s - p_n)}. \tag{4.7}$$

We assume that all the poles are simple ones for simplicity. Note that $m \le n$ holds for a causal system. Let the input of the system be

$$u(t) = A \sin \omega t. \tag{4.8}$$

The Laplace transform of (4.8) (see Table 2.1) is

$$U(s) = \frac{A\omega}{s^2 + \omega^2} = \frac{A\omega}{(s + j\omega)(s - j\omega)}. \tag{4.9}$$

The output of the system is

$$\begin{aligned} Y(s) &= \frac{b_0 s^m + b_1 s^{m-1} + \cdots + b_{m-1}s + b_m}{(s - p_1)(s - p_2) \cdots (s - p_n)} \frac{A\omega}{(s + j\omega)(s - j\omega)} \\ &= A\left\{ \frac{G(j\omega)}{2j(s - j\omega)} - \frac{G(-j\omega)}{2j(s + j\omega)} \right\} + \sum_{i=1}^{n} \frac{c_i}{s - p_i}, \end{aligned} \tag{4.10}$$

where

$$c_i = (s - p_i)G(s)\frac{A\omega}{s^2 + \omega^2}\bigg|_{s=p_i} , \quad i = 1, 2, \cdots, n.$$

Thus, $y(t) = \mathcal{L}^{-1}\{Y(s)\}$ is given by

$$\begin{aligned} y(t) &= A|G(j\omega)|\frac{e^{j(\omega t + \phi)} - e^{-j(\omega t + \phi)}}{2j} + \sum_{i=1}^{n} c_i e^{p_i t} \\ &= A|G(j\omega)|\sin(\omega t + \phi) + \sum_{i=1}^{n} c_i e^{p_i t}, \end{aligned} \tag{4.11}$$

where $\phi = \angle G(j\omega)$ (It is explained in the next section.). If $p_i < 0$ ($i = 1, 2, \cdots, n$), then $\lim_{t \to \infty} \sum_{i=1}^{n} c_i e^{p_i t} \to 0$. That is, if the system is stable, the output is a sine wave with the same angular frequency of the input and is $|G(j\omega)|$ times larger than the input when the time is sufficiently long. The difference between the phases of the output and input is ϕ. It is *called phase lead* (位相進み) if $\phi > 0$ and *phase lag* (位相遅れ) if $\phi < 0$. The above shows that $G(j\omega)$ $[|G(j\omega)|$ and $\angle G(j\omega)]$ plays an important role in system response.

4.3   Nyquist Plot

A *Nyquist plot* (or a *polar plot*) (ナイキスト軌跡) is a technique that investigates the characteristics of $G(j\omega)$ [we obtain it by letting $s = j\omega$ for $G(s)$] visually in Cartesian coordinates. First, we write $G(j\omega)$ as

$$G(j\omega) = X(\omega) + jY(\omega), \tag{4.12}$$

where $X(\omega)$ is the real part and $Y(\omega)$ is the imaginary part. Note that a given ω corresponds to[†1] a point, $(X(\omega),\ Y(\omega))$ in the s plane. The amplitude and phase of $G(j\omega)$ are

$$|G(j\omega)| = \sqrt{X^2(\omega) + Y^2(\omega)}, \ \angle G(j\omega) = \tan^{-1}\frac{Y(\omega)}{X(\omega)}. \tag{4.13}$$

Plotting the points from $\omega = -\infty$ to $\omega = +\infty$ gives the Nyquist locus of the system. Moreover, since the locus from $\omega = 0$ to $\omega = -\infty$ and that from $\omega = 0$ to $\omega = +\infty$ are symmetric to the x-axis, we only need to draw the part from $\omega = 0$ to $\omega = +\infty$.

The Nyquist stability criterion (it is explained in Chapter 5) uses the Nyquist plot to determine whether or not[†2] a system is stable. Not only the stability of a *closed-loop system* (閉ループシステム) but also its relative stability can easily be checked from the frequency response of its *open-loop system* (開ループシステム).

【**Example 4.2**】   Draw the Nyquist plot for a *first-order system* (1 次系)

$$G_1(s) = \frac{1}{Ts+1}. \tag{4.14}$$

Note

[†1]   correspond to ～：～に対応する, ～に相当する。
[†2]   whether or not ～：～かどうか。

$$\begin{cases} G_1(j\omega) = \dfrac{1}{Tj\omega + 1} = \dfrac{1}{1 + (T\omega)^2} - j\dfrac{T\omega}{1 + (T\omega)^2}, \\ |G_1(j\omega)| = \dfrac{1}{\sqrt{1 + (T\omega)^2}}, \quad \angle G_1(j\omega) = -\tan^{-1} T\omega. \end{cases} \tag{4.15}$$

The real and imaginary parts are

$$\begin{cases} X(\omega) = \mathrm{Re}[G_1(j\omega)] = \dfrac{1}{1 + (T\omega)^2}, \\ Y(\omega) = \mathrm{Im}[G_1(j\omega)] = -\dfrac{T\omega}{1 + (T\omega)^2}. \end{cases} \tag{4.16}$$

Calculating several points (**Table 4.1**), we connect these points and draw the Nyquist plot as **Figure 4.6**.

Table 4.1   Frequency response for (4.14)

Point	$T\omega$	$G_1(j\omega)$	$|G_1(j\omega)|$	$\angle G_1(j\omega)$
A	0	$1 - j0$	1	$0°$
B	0.42	$0.85 - j0.36$	0.92	$-22.78°$
C	0.73	$0.65 - j0.48$	0.81	$-36.13°$
D	1	$0.5 - j0.5$	0.71	$-45°$
E	1.36	$0.35 - j0.48$	0.59	$-53.67°$
F	2.38	$0.15 - j0.36$	0.39	$-67.21°$
H	10	$9.9 \times 10^{-3} - j0.1$	0.1	$-84.29°$

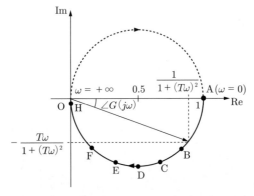

Figure 4.6   Nyquist plot for (4.15)

On the other hand, it is easy to check that

$$\left[X(\omega) - \frac{1}{2}\right]^2 + Y^2(\omega) = \left(\frac{1}{2}\right)^2. \tag{4.17}$$

So, the trajectory of $G_1(j\omega)$ is a circle with its center point being $(1/2, 0)$.

The following MATLAB commands compute the Nyquist plot of (4.14) for $T = 1$:

Program 4.1   Nyquist plot for (4.14)

```
1  s=tf('s');
2  T=1; G1=1/(T*s+1)
3  nyquistplot(G1);
4  axis([-1.5 1.5 -1 1]);
```

4.4   Bode Plots

As shown in the previous section, the relationship between $G(j\omega)$ and ω is nonlinear. It is not easy to draw a Nyquist plot. Moreover, a typical feedback control system contains a plant $[P(s)]$ and a controller $[C(s)]$ (Figure 1.7 in Chapter 1). Even though we have the Nyquist plots of $P(s)$ and $C(s)$, we cannot directly obtain the Nyquist plot of $L(s) = C(s)P(s)$ and have to redraw $L(j\omega)$ from the beginning. Another problem with a Nyquist plot is that it is hard to obtain precise information of angular frequencies from the plot.

Bode plots (ボード線図) provides us with a way to solve these problems. Since $\log AB = \log A + \log B$, if we know the Bode plots of $P(s)$ and $C(s)$, simply adding them together gives that of $L(s)$. This makes it an easy task to draw Bode plots. While a Bode phase plot is nonlinear, a Bode magnitude plot can use straight lines to approximate an original curve. It uses only a couple of simple rules to easily draw the plot. Thus, Bode plots are widely used to analyze the frequency response of a linear time-invariant system.

We can describe $G(j\omega)$ as

$$G(j\omega) = |G(j\omega)|e^{j\phi(\omega)}, \ \phi(\omega) = \angle G(j\omega), \tag{4.18}$$

where $|G(j\omega)|$ and $\angle G(j\omega)$ are the magnitude and the phase shift of the frequency response, respectively.

We use a logarithm to describe the gain characteristic

$$g(\omega) = 20\log_{10}|G(j\omega)| \text{ [dB]}. \tag{4.19}$$

A Bode magnitude plot is drawn as a linear scale in decibels for the vertical axis and a logarithmic scale of angular frequency in radian per second for the horizontal axis; and a Bode phase plot, a linear scale in degree for the vertical axis and the same as the magnitude plot for the horizontal axis.

【**Example 4.3**】 Draw Bode plots for a *proportional element* (比例要素)

$$G_K(s) = K. \tag{4.20}$$

Since

$$g_K(\omega) = 20\log_{10} K, \ \angle G_K(j\omega) = \tan^{-1} 0 = 0, \tag{4.21}$$

Bode plots for $K = 100$ in (4.20) are shown in **Figure 4.7**.

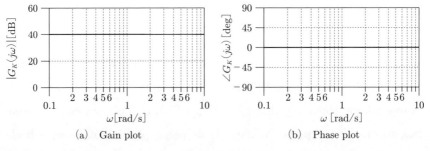

(a) Gain plot   (b) Phase plot

Figure 4.7 Bode plots of $G_K(s)$ for $K = 100$ in (4.20)

【**Example 4.4**】 Draw Bode plots for a *time-delay element* (時間遅れ
要素)

$$G_d(s) = e^{-Ts}, \tag{4.22}$$

where T is the delay time.

Since

$$g_d(\omega) = 20\log_{10}|e^{-jT\omega}| = -20\log_{10} 1 = 0, \quad \angle G_d(j\omega) = -T\omega, \tag{4.23}$$

Bode plots for $T = 1$ s in (4.22) are shown in **Figure 4.8**.

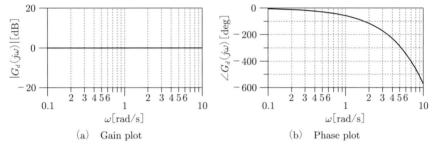

(a)   Gain plot   (b)   Phase plot

Figure 4.8 Bode plots of $G_d(s)$ for $T = 1$ in (4.22)

$G_d(s)$ is not a rational function but a transcendental one. Since its gain
is one but its phase lag is in proportion to[†] the angular frequency, it is
difficult to handle this element in control-system design. Note that

$$G_d(s) = e^{-Ts} = 1 - Ts + \cdots + (-1)^n\frac{T^n}{n!}s^n + \cdots, \tag{4.24}$$

$$G_d(s) = \frac{1}{e^{Ts}} = \frac{1}{1 + Ts + \cdots + \dfrac{T^n}{n!}s^n + \cdots}, \tag{4.25}$$

$$G_d(s) = \frac{e^{-Ts/2}}{e^{Ts/2}} = \frac{1 - \dfrac{T}{2}s + \cdots + (-1)^n\dfrac{1}{n!}\left(\dfrac{T}{2}\right)^n s^n + \cdots}{1 + \dfrac{T}{2}s + \cdots + \dfrac{1}{n!}\left(\dfrac{T}{2}\right)^n s^n + \cdots}. \tag{4.26}$$

[†]   in proportion to 〜：〜に正比例している，〜にしたがって。

We can consider that $G_d(s)$ contains infinity numbers of poles and zeros. This makes system design a difficult task. Padé approximation of a time delay is often used to derive a rational model for system design. **Table 4.2** shows the approximation for the degrees of the denominator and numerator polynomials being n and m, respectively.

Table 4.2 Padé approximation of (4.22)

m \ n	1	2	3
0	$\dfrac{1}{1+Ts}$	$\dfrac{1}{1+Ts+\dfrac{T^2}{2}s^2}$	$\dfrac{1}{1+Ts+\dfrac{T^2}{2}s^2+\dfrac{T^3}{6}s^3}$
1	$\dfrac{1-\dfrac{T}{2}s}{1+\dfrac{T}{2}s}$	$\dfrac{1-\dfrac{T}{3}s}{1+\dfrac{2T}{3}s+\dfrac{T^2}{6}s^2}$	$\dfrac{1-\dfrac{T}{4}s}{1+\dfrac{3T}{4}s+\dfrac{T^2}{4}s^2+\dfrac{T^3}{24}s^3}$
2		$\dfrac{1-\dfrac{T}{2}s+\dfrac{T^2}{12}s^2}{1+\dfrac{T}{2}s+\dfrac{T^2}{12}s^2}$	$\dfrac{1-\dfrac{3T}{5}s+\dfrac{T^2}{20}s^2}{1+\dfrac{3T}{5}s+\dfrac{3T^2}{20}s^2+\dfrac{T^3}{60}s^3}$
3			$\dfrac{1-\dfrac{T}{2}s+\dfrac{T^2}{10}s^2-\dfrac{T^3}{120}s^3}{1+\dfrac{T}{2}s+\dfrac{T^2}{10}s^2+\dfrac{T^3}{120}s^3}$

【**Example 4.5**】 Draw Bode plots for an *integral element* (積分要素)

$$G_I(s) = \frac{1}{s}. \tag{4.27}$$

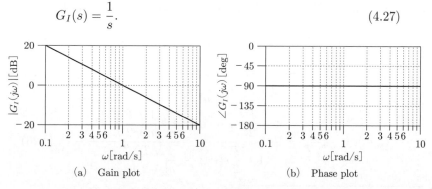

(a) Gain plot

(b) Phase plot

Figure 4.9 Bode plots of $G_I(s)$ in (4.27)

Since

$$g_I(\omega) = 20\log_{10}\frac{1}{\omega} = -20\log_{10}\omega, \ \angle G_I(j\omega) = -90^\circ, \quad (4.28)$$

Bode plots for (4.27) are shown in **Figure 4.9**.

【**Example 4.6**】 Draw Bode plots for a *differential element* (微分要素)

$$G_D(s) = s. \tag{4.29}$$

Since

$$g_D(\omega) = 20\log_{10}\omega = 20\log_{10}\omega, \ \angle G_D(j\omega) = 90^\circ, \tag{4.30}$$

Bode plots for (4.29) are shown in **Figure 4.10**.

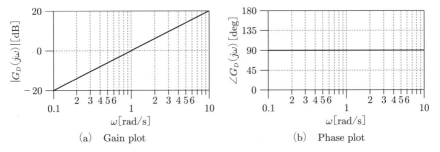

(a) Gain plot   (b) Phase plot

Figure 4.10 Bode plots of $G_D(s)$ in (4.29)

【**Example 4.7**】 Draw Bode plots for a first-order system (4.14).

Replacing s with jw in (4.14) and calculating its gain and phase yield (4.15). Thus,

$$g_1(\omega) = 20\log_{10}|G_1(j\omega)| = -20\log_{10}\sqrt{1 + (T\omega)^2}. \tag{4.31}$$

For $T\omega \ll 1$ ($\omega \ll 1/T$),

$$g_1(\omega) \fallingdotseq -20\log_{10}1 = 0 \text{ dB}. \tag{4.32}$$

For $T\omega \gg 1$ $(\omega \gg 1/T)$,

$$g_1(\omega) \doteqdot -20\log_{10} T\omega. \tag{4.33}$$

More specifically, for $\omega_1 = 10\omega_0$, (4.33) yields

$$g_1(\omega_1) \doteqdot -20\log_{10} T(10\omega_0) = -20 - 20\log_{10} T\omega_0$$
$$= -20 + g_1(\omega_0). \tag{4.34}$$

Thus, the gain decreases $-20\,\text{dB}$ as the angular frequency increases 10 times. We use $-20\,\text{dB/dec}^{\dagger}$ in the plot to indicate this.

As a result, we can use two asymptotes to simply draw the magnitude plot: Draw a line of $0\,\text{dB}$ from a small ω up to $\omega = 1/T$. Then, draw a line that passes two points: $(1/T, 0)$ and $(10/T, -20)$. These two asymptotes cross at the angular frequency $\omega = 1/T$, which is called the breakpoint. The biggest error between the piecewise linear approximations and the actual magnitude curve occurs at $\omega = 1/T$. It is $-3\,\text{dB}$ $\left(= 20\log_{10} 1/\sqrt{2}\right)$.

We can also use asymptotes to draw the Bode phase plot

$$\angle G_1(j\omega) = -\tan^{-1} T\omega. \tag{4.35}$$

Note that, for $T\omega \ll 1$ $(\omega \ll 1/T)$,

$$\angle G_1(j\omega) \doteqdot 0°. \tag{4.36}$$

For $T\omega \gg 1$ $(\omega \gg 1/T)$,

$$\angle G_1(j\omega) \doteqdot -90°. \tag{4.37}$$

For $T\omega = 1$ $(\omega = 1/T)$,

$$\angle G_1(j\omega) = -45°. \tag{4.38}$$

$\dagger$　dec：decade の略，10 倍。

So, we can use three asymptotes to draw the phase plot: a line of $0°$ from a small ω up to $\omega = 1/(5T)$, a line of $-90°$ for angular frequencies equal or higher than $\omega = 5/T$, and a line that passes three points: $(1/(5T), 0°)$, $(1/T, -45°)$, and $(5/T, -90°)$. The biggest error between the asymptotes and the actual phase curve is $11°$ (at $\omega T = 0.2$ and $\omega T = 5$), which is relatively big. We usually use **Table 4.3** to precisely draw the phase plot.

Table 4.3   Inverse tangent

ωT	$\tan^{-1}\omega T$	ωT	$\tan^{-1}\omega T$	ωT	$\tan^{-1}\omega T$
0.1	5.71°	0.7	35.0°	5	78.7°
0.2	11.3°	0.8	38.7°	6	80.5°
0.3	16.7°	0.9	42.0°	7	81.9°
0.4	21.8°	2	63.4°	8	82.9°
0.5	26.6°	3	71.6°	9	83.7°
0.6	31.0°	4	76.0°	10	84.3°

As a result, Bode plots for the system (4.14) are shown in **Figure 4.11**.

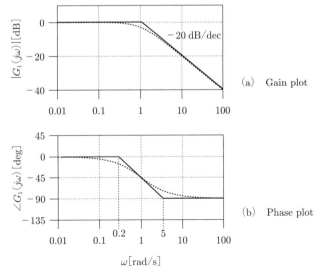

Figure 4.11   Bode plots of $G_1(s)$ for $T = 1$ in (4.14)

【**Example 4.8**】 Draw Bode plots for a *proportional-differential element*
(比例微分要素)

$$G_{1D}(s) = Ts + 1. \tag{4.39}$$

Since

$$\begin{cases} G_{1D}(j\omega) = Tj\omega + 1, \\ |G_{1D}(j\omega)| = \sqrt{1 + (T\omega)^2}, \ \angle G_{1D}(j\omega) = \tan^{-1} T\omega, \end{cases} \tag{4.40}$$

we have

$$\begin{cases} g_{1D}(\omega) = 20\log_{10}|G_{1D}(j\omega)| = 20\log_{10}\sqrt{1 + (T\omega)^2}, \\ \angle G_{1D}(j\omega) = \tan^{-1} T\omega. \end{cases} \tag{4.41}$$

Comparing (4.41) with (4.31) and (4.35) shows that the magnitude plot for (4.39) and that for the first-order system (4.14) are symmetric to the horizontal axis of 0 dB; the phase plot for (4.39) and that for (4.35), the horizontal axis of 0°. The plots are shown in **Figure 4.12**.

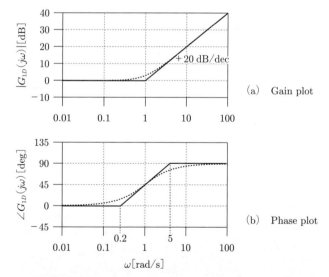

(a) Gain plot

(b) Phase plot

Figure 4.12 Bode plots of $G_{1D}(s)$ for $T = 1$ in (4.39)

【Example 4.9】 Draw Bode plots for a *second-order system* (2 次系) and its *inverse system* (逆システム)

$$G_2(s) = \frac{\omega_n^2}{s^2 + 2\zeta\omega_n s + \omega_n^2}, \quad G_{2D}(s) = \frac{s^2 + 2\zeta\omega_n s + \omega_n^2}{\omega_n^2}. \tag{4.42}$$

Calculating the gains and phases of (4.42) yields

$$\begin{cases} g_2(\omega) = -20\log_{10}\sqrt{\left[1 - \left(\frac{\omega}{\omega_n}\right)^2\right]^2 + 4\zeta^2\left(\frac{\omega}{\omega_n}\right)^2}, \\[3mm] \angle G_2(j\omega) = -\tan^{-1}\frac{2\zeta(\omega/\omega_n)}{1 - (\omega/\omega_n)^2}, \\[3mm] g_{2D}(\omega) = 20\log_{10}\sqrt{\left[1 - \left(\frac{\omega}{\omega_n}\right)^2\right]^2 + 4\zeta^2\left(\frac{\omega}{\omega_n}\right)^2}, \\[3mm] \angle G_{2D}(j\omega) = \tan^{-1}\frac{2\zeta(\omega/\omega_n)}{1 - (\omega/\omega_n)^2}. \end{cases} \tag{4.43}$$

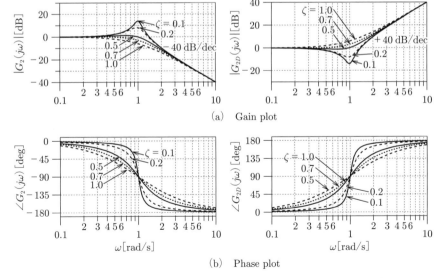

(a)   Gain plot

(b)   Phase plot

Figure 4.13   Bode plots of $G_2(s)$ and $G_{2D}(s)$ for $\omega_n = 1$ in (4.42)

Clearly, Bode plots for $G_2(s)$ and $G_{2D}(s)$ are symmetric to the horizontal axis of 0 dB for the magnitude and to the horizontal axis of $0°$ for the phase (**Figure 4.13**). For $G_2(s)$, a simple calculation gives

$$\frac{\omega}{\omega_n} \ll 1 \ (\omega \ll \omega_n)\text{: } g_2(\omega) \doteq 0 \text{ dB and } \angle G_2(j\omega) \doteq 0°;$$

$$\frac{\omega}{\omega_n} = 1 \ (\omega = \omega_n)\text{: } g_2(\omega) = -20\log_{10} 2\zeta \text{ dB and } \angle G_2(j\omega) = -90°;$$

$$\frac{\omega}{\omega_n} \gg 1 \ (\omega \gg \omega_n)\text{: } g_2(\omega) \doteq -40\log_{10} \frac{\omega}{\omega_n} \text{ dB and } \angle G_2(\omega) \doteq -180°.$$

In general[†], a transfer function is

$$G(s) = \frac{Ke^{-Ts} \prod\limits_{k=1}^{m_1}(1 + T_{Dk}s) \prod\limits_{q=1}^{m_2}(s^2 + 2\zeta_{Dq}\omega_{Dq}s + \omega_{Dq}^2)}{s^l \prod\limits_{j=1}^{p_1}(1 + T_j s) \prod\limits_{n=1}^{p_2}(s^2 + 2\zeta_n\omega_n s + \omega_n^2)}, \quad (4.44)$$

that is, it is a combination of (4.14), (4.20), (4.22), (4.27), (4.39), and (4.42). Both the magnitude and phase of (4.44) are plotted by simply adding the terms graphically. This makes it an easy task to draw Bode plots. For example,

$$\begin{aligned} G(s) &= \frac{K(1 + T_Ds)(s^2 + 2\zeta_D\omega_D s + \omega_D^2)}{s(1 + Ts)(s^2 + 2\zeta_n\omega_n s + \omega_n^2)} \\ &= \frac{K\omega_D^2}{\omega_n^2} \times \frac{1}{s} \times \frac{1}{1 + Ts} \times (1 + T_Ds) \\ &\quad \times \frac{\omega_n^2}{s^2 + 2\zeta_n\omega_n s + \omega_n^2} \times \frac{s^2 + 2\zeta_D\omega_D s + \omega_D^2}{\omega_D^2} \\ &= |G_K(\omega)|e^{j\phi_K(\omega)} \cdot |G_I(\omega)|e^{j\phi_I(\omega)} \cdot |G_1(\omega)|e^{j\phi_1(\omega)} \\ &\quad \cdot |G_{1D}(\omega)|e^{j\phi_{1D}(\omega)} \cdot |G_2(\omega)|e^{j\phi_2(\omega)} \cdot |G_{2D}(\omega)|e^{j\phi_{2D}(\omega)}, \end{aligned} \quad (4.45)$$

and

$$g_K(\omega) = 20\log_{10}|G_K(j\omega)| = 20\log_{10} \frac{K\omega_D^2}{\omega_n^2}, \quad \phi_K(\omega) = 0,$$

---

† in general：一般に。

$$g_I(\omega) = 20\log_{10}|G_I(j\omega)| = -20\log_{10}\omega, \ \ \phi_I(\omega) = -90,$$

$$g_1(\omega) = 20\log_{10}|G_1(j\omega)| = -20\log_{10}\sqrt{1+(T\omega)^2},$$

$$\phi_1(\omega) = -\tan^{-1}T\omega,$$

$$g_{1D}(\omega) = 20\log_{10}|G_{D1}(j\omega)| = 20\log_{10}\sqrt{1+(T_D\omega)^2},$$

$$\phi_{1D}(\omega) = \tan^{-1}T_D\omega,$$

$$g_2(\omega) = 20\log_{10}|G_2(j\omega)| = -20\log_{10}\sqrt{\left[1-\left(\frac{\omega}{\omega_n}\right)^2\right]^2 + 4\zeta^2\left(\frac{\omega}{\omega_n}\right)^2},$$

$$\phi_2(\omega) = -\tan^{-1}\frac{2\zeta(\omega/\omega_n)}{1-(\omega/\omega_n)^2},$$

$$g_{2D}(\omega) = 20\log_{10}|G_{2D}(j\omega)| = 20\log_{10}\sqrt{\left[1-\left(\frac{\omega}{\omega_D}\right)^2\right]^2 + 4\zeta_D^2\left(\frac{\omega}{\omega_D}\right)^2},$$

$$\phi_{2D}(\omega) = \tan^{-1}\frac{2\zeta_D(\omega/\omega_D)}{1-(\omega/\omega_D)^2}.$$

The Bode magnitude plot is

$$g(\omega) = g_K(\omega) + g_I(\omega) + g_1(\omega) + g_{1D}(\omega) + g_2(\omega) + g_{2D}(\omega), \quad (4.46)$$

and the Bode phase plot is

$$\phi(\omega) = \phi_K(\omega) + \phi_I(\omega) + \phi_1(\omega) + \phi_{1D}(\omega) + \phi_2(\omega) + \phi_{2D}(\omega). \quad (4.47)$$

A linear, time-invariant system is called a *minimum-phase system* (最小位相系) if all the poles and zeros of the system are stable (A zero is said to be stable if it is in the left-half s plane.). Otherwise, it is called a *nonminimum-phase system* (非最小位相系). The Bode magnitude and phase plots of a minimum-phase system have the same trends, that is, the phase increases as the magnitude increases.

————— **Problems** —————

⟨ **Basic Level** ⟩

[1] Find transfer functions of systems in **Figure 4.14**.

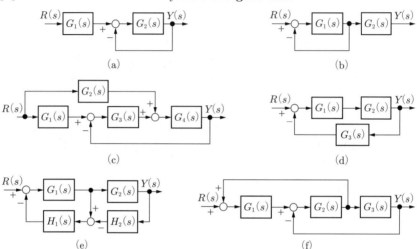

Figure 4.14   Problem 1

[2] Find the transfer functions of a system in **Figure 4.15**:

$$\begin{bmatrix} Y_1(s) \\ Y_2(s) \end{bmatrix} = \begin{bmatrix} G_{11}(s) & G_{12}(s) \\ G_{21}(s) & G_{22}(s) \end{bmatrix} \begin{bmatrix} R_1(s) \\ R_2(s) \end{bmatrix}.$$

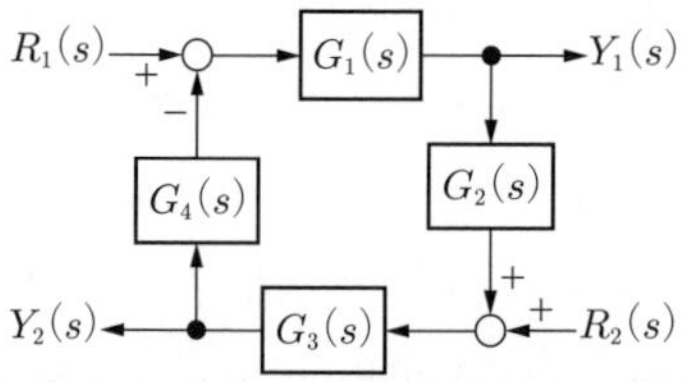

Figure 4.15   Problem 2

[3] Draw a Nyquist plot for an open-loop transfer function $L(s) = C(s)P(s) = 10e^{-Ts}/[s(s+1)]$ in a unity-feedback system and use MATLAB to verify them:

   A. $T = 0\,\text{s}$ B. $T = 0.1\,\text{s}$ C. $T = 3\,\text{s}$.

[4] Draw Bode plots for open-loop transfer functions given in Problem 3 and use MATLAB to verify them.

[5] Draw Nyquist plots for the following open-loop transfer functions and use MATLAB to verify them:

A. $L(s) = \dfrac{1}{s(s+1)}$

B. $L(s) = \dfrac{1}{(s+1)(5s+1)}$

C. $L(s) = \dfrac{2}{s^2(s+1)}$

D. $L(s) = \dfrac{3(s+2)}{s(s-1)}$

E. $L(s) = \dfrac{s+2}{(s+1)(0.1s+1)}$

F. $L(s) = \dfrac{s-2}{(s+1)(0.1s+1)}.$

[6] Draw Bode plots for open-loop transfer functions given in Problem 5 and use MATLAB to verify them.

[7] Draw Bode plots for the following open-loop transfer functions and use MATLAB to verify them:

A. $L(s) = \dfrac{10}{(s+1)(10s+1)}$

B. $L(s) = \dfrac{2}{(2s+1)(5s+1)}$

C. $L(s) = \dfrac{200}{s^2(s+1)(10s+1)}$

D. $L(s) = \dfrac{8(10s+1)}{s(0.5s+1)(s^2+s+1)}$

E. $L(s) = \dfrac{4\,000\left(s^2+40s+400\right)}{s(s+1)(10s+1)}$

F. $L(s) = \dfrac{s+1}{9s(0.5s+1)\left(s^2+3s+9\right)}.$

Advanced Level

[1] Find the input-output transfer functions of systems in **Figure 4.16**.

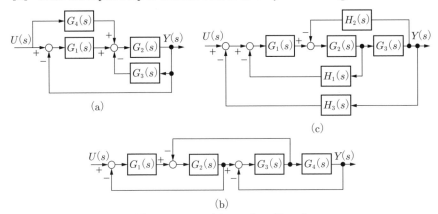

Figure 4.16   Advanced problem 1

[2] Assume that an open-loop transfer function, $L(s)$, is minimum-phase. For the piecewise linear asymptotic Bode magnitude plots for $L(s)$ in **Figure 4.17**,

    A.   find the transfer functions of $L(s)$,

    B.   draw a Nyquist plot for $L(s)$.

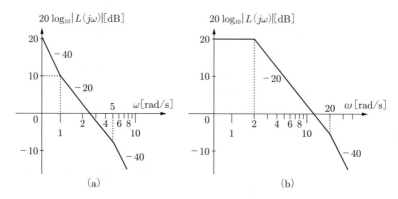

Figure 4.17 Advanced problem 2

[3] Draw Bode plots for $G_m(s) = (1+s)/(s^2 + s + 1)$ and $G_n(s) = (1-s)/(s^2 + s + 1)$ observe the similarities and differences in the plots for these two systems.

5 | Stability Analysis

In the design of a control system, we need to find a controller that ensures the performance of a closed-loop control system. To begin with[†], we have to guarantee the stability of the system.

The response of a system is usually divided into transient and steady-state responses. If the transient response of a system gradually decays and finally converges to a steady state, we say that the system is *stable* (安定). However, if the transient response diverges, we say that the system is *unstable* (不安定).

In this chapter, we first observe the relationship between the stability and the poles for first- and second-order systems. Then, we explain three stability criteria based on such a relationship.

5.1 First-Order System

Consider a first-order system

$$T\frac{dy(t)}{dt} + y(t) = Ku(t), \tag{5.1}$$

where $u(t)$ is the input and $y(t)$ is the output of the system. Taking the Laplace transform of (5.1) for $y(0) = 0$ yields

$$TsY(s) + Y(s) = KU(s). \tag{5.2}$$

Thus, we obtain the transfer function from the input to the output

$$G(s) = \frac{Y(s)}{U(s)} = \frac{K}{Ts+1}. \tag{5.3}$$

[†] to begin with：まず第一に，最初に。

Note that

$$Y(s) = \frac{K}{Ts+1}U(s), \tag{5.4}$$

it is easy to use the inverse Laplace transform to calculate the output of the system. For example, since $\mathcal{L}\{\delta(t)\} = 1$, the impulse response of the system is

$$y(t) = \mathcal{L}^{-1}\left\{\frac{K}{Ts+1} \times 1\right\} = \frac{K}{T}e^{-\frac{t}{T}}, \tag{5.5}$$

and since $\mathcal{L}\{1(t)\} = 1/s$, the step response of the system is

$$y(t) = \mathcal{L}^{-1}\left\{\frac{K}{Ts+1} \times \frac{1}{s}\right\} = \mathcal{L}^{-1}\left\{\frac{K}{s} - \frac{K}{s+1/T}\right\} = K\left(1 - e^{-t/T}\right). \tag{5.6}$$

On the other hand,

$$G(s) = \frac{N(s)}{D(s)}, \ N(s) = K, \ D(s) = Ts+1. \tag{5.7}$$

The pole of the system is $p = -1/T$.

The impulse and step responses of the system (5.1) are calculated by the following MATLAB commands:

```
Program 5.1   Impulse/step resp. of system (5.1)
 1  s=tf('s'); % Laplace operator
 2  t=0:0.01:2; % Time vector for simulation (stable)
 3  tn=0:0.01:1; % Time vector for simulation (unstable)
 4  K=1; % DC gain
 5  T=0.5; % Time constant
 6  G05=K/(T*s+1) % First-order system
 7  yi05=impulse(G05,t); %Impulse response
 8  ys05=step(G05,t); % Step response
 9
10  T=0.2; G02=K/(T*s+1)
11  yi02=impulse(G02,t); ys02=step(G02,t);
```

```
12
13  T=0.1; G01=K/(T*s+1)
14  yi01=impulse(G01,t); ys01=step(G01,t);
15
16  T=-0.2; Gn02=K/(T*s+1)
17  yin02=impulse(Gn02,tn); ysn02=step(Gn02,tn);
18
19  T=-0.5; Gn05=K/(T*s+1)
20  yin05=impulse(Gn05,tn); ysn05=step(Gn05,tn);
21
22  figure(1);
23  plot(t,yi05,t,yi02,t,yi01);
24  grid; % Draw grids
25
26  figure(2);
27  plot(t,ys05,t,ys02,t,ys01); grid;
28
29  figure(3);
30  plot(tn,yin02,tn,yin05); grid;
31
32  figure(4);
33  plot(tn,ysn02,tn,ysn05); grid;
```

The time responses (**Figure 5.1**) show that

i)   the system is stable if the pole is in the open left-half s plane (that is, $T > 0$);

ii) the system is unstable if the pole is in the open right-half s plane (that is, $T < 0$); and

iii) the system decays or diverges faster as the pole is located farther from the origin.

There are two parameters in (5.1) and (5.3): K is a *DC* (*direct current*) *gain* (直流ゲイン) and T is a *time constant* (時定数). K shows how many

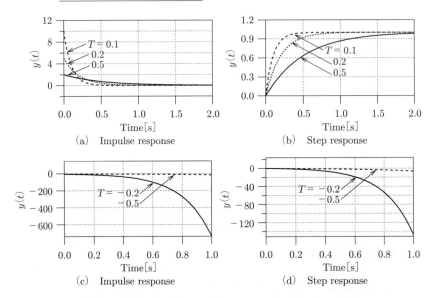

Figure 5.1 Impulse and step responses of system (5.1)

times the D.C. component of an input signal is amplified by the system. T means that the output of the system has changed about 63% toward its final value in one time constant. For example, the step response of the system reaches 0.632 $(= 1 - e^{-1})$ of its final value when $t = T$. Moreover, it is 0.865 when $t = 2T$, 0.950 when $t = 3T$, and 0.982 when $t = 4T$. Thus, T is a parameter governing the response speed. A small T (thus a pole far from the origin) results in[†] a fast response and vice versa.

5.2   Second-Order System

The standard form of a second-order system is

$$G(s) = \frac{\omega_n^2}{s^2 + 2\zeta\omega_n s + \omega_n^2},\tag{5.8}$$

where ζ is the *damping ratio* (減衰係数，減衰定数，減衰比) and ω_n is the

---

[†]   result in ～：～という結果になる，～という結果をもたらす。

natural angular frequency (固有角周波数, 自然角周波数) of the system. The poles of the system are given by

$$D(s) = s^2 + 2\zeta\omega_n s + \omega_n^2 = 0.$$

A simple calculation yields

$$p_{1,2} = -\zeta\omega_n \pm \sqrt{\zeta^2 - 1}\omega_n. \tag{5.9}$$

The impulse response is as follows:

(**1**)   When $p_1 \neq p_2$ and p_1 and p_2 are real roots ($\zeta > 1$),

$$\begin{aligned} y(t) &= \mathcal{L}^{-1}\left\{\frac{\omega_n^2}{(s - p_1)(s - p_2)}\right\} = \mathcal{L}^{-1}\left\{\frac{A_1}{s - p_1} + \frac{A_2}{s - p_2}\right\} \\ &= A_1 e^{p_1 t} + A_2 e^{p_2 t}, \\ &= \frac{\omega_n}{2\sqrt{\zeta^2 - 1}}\left[e^{-(\zeta - \sqrt{\zeta^2 - 1})\omega_n t} - e^{-(\zeta + \sqrt{\zeta^2 - 1})\omega_n t}\right]; \end{aligned} \tag{5.10}$$

(**2**)   When $p_1 = p_2$ ($\zeta = 1$), that is, a double real root,

$$y(t) = \mathcal{L}^{-1}\left\{\frac{\omega_n^2}{(s - p_1)^2}\right\} = \omega_n^2 t e^{-\omega_n t}, \tag{5.11}$$

which is obtained from Table 2.1;

(**3**)   When $p_1 \neq p_2$ and they are a pair of complex conjugate roots ($0 \leq \zeta < 1$),

$$\begin{aligned} y(t) &= \mathcal{L}^{-1}\left\{\frac{\omega_n^2}{(s + \zeta\omega_n)^2 + (\sqrt{1 - \zeta^2}\omega_n)^2}\right\} \\ &= \frac{\omega_n}{\sqrt{1 - \zeta^2}}e^{-\zeta\omega_n t}\sin\omega_d t, \end{aligned} \tag{5.12}$$

where

$$\omega_d = \sqrt{1 - \zeta^2}\omega_n. \tag{5.13}$$

The step response is as follows:

(**1**)   When $p_1 \neq p_2$ and p_1 and p_2 are real roots ($\zeta > 1$),

$$y(t) = \mathcal{L}^{-1}\left\{\frac{\omega_n^2}{(s-p_1)(s-p_2)} \cdot \frac{1}{s}\right\} = \mathcal{L}^{-1}\left\{\frac{A_0}{s} + \frac{A_1}{s-p_1} + \frac{A_2}{s-p_2}\right\}$$

$$= A_0 + A_1 e^{p_1 t} + A_2 e^{p_2 t},$$

$$= 1 - \frac{\zeta+\sqrt{\zeta^2-1}}{2\sqrt{\zeta^2-1}}e^{-(\zeta-\sqrt{\zeta^2-1})\omega_n t} + \frac{\zeta-\sqrt{\zeta^2-1}}{2\sqrt{\zeta^2-1}}e^{-(\zeta+\sqrt{\zeta^2-1})\omega_n t};$$

$$(5.14)$$

(2) When $p_1 = p_2$ ($\zeta = 1$),

$$y(t) = \mathcal{L}^{-1}\left\{\frac{\omega_n^2}{(s-p_1)^2} \cdot \frac{1}{s}\right\} = \mathcal{L}^{-1}\left\{\frac{1}{s} - \frac{\omega_n}{(s+\omega_n)^2} - \frac{1}{s+\omega_n}\right\}$$

$$= 1 - (1 + \omega_n t)e^{-\omega_n t}. \qquad (5.15)$$

(3) When $p_1 \neq p_2$ and they are a pair of complex conjugate roots $(0 \leq \zeta < 1)$,

$$y(t) = \mathcal{L}^{-1}\left\{\frac{\omega_n^2}{(s+\zeta\omega_n)^2 + (\sqrt{1-\zeta^2}\omega_n)^2} \cdot \frac{1}{s}\right\}$$

$$= \mathcal{L}^{-1}\left\{\frac{1}{s} - \frac{s+2\zeta\omega_n}{(s+\zeta\omega_n)^2 + (\sqrt{1-\zeta^2}\omega_n)^2}\right\}$$

$$= 1 - \frac{1}{\sqrt{1-\zeta^2}}e^{-\zeta\omega_n t}\sin(\omega_d t + \phi), \qquad (5.16)$$

where ω_d is given in (5.13) and $\phi = \tan^{-1}(\sqrt{1-\zeta^2}/\zeta)$.

The impulse and step responses of the system (5.8) are calculated by the following MATLAB commands:

```
Program 5.2   Impulse/step resp. of system (5.8)

1 s=tf('s'); % Laplace operator
2 t=0:0.01:50;
3
4 wn=1; % natural angular frequency
5 zai=0.1; % Damping ratio
6 G01=wn^2/(s^2+2*wn*zai*s+wn^2)
7 yi01=impulse(G01,t); ys01=step(G01,t);
8
```

```
9  zai=0.5; G05=wn^2/(s^2+2*+wn*zai*s+wn^2)
10 yi05=impulse(G05,t); ys05=step(G05,t);
11
12 zai=0.7; G07=wn^2/(s^2+2*wn*zai*s+wn^2)
13 yi07=impulse(G07,t); ys07=step(G07,t);
14
15 figure(1);
16 plot(t,yi01,t,yi05,t,yi07); grid;
17
18 figure(2);
19 plot(t,ys01,t,ys05,t,ys07); grid;
```

Note that the gradients of the complex conjugate poles are
$K_\zeta = \dfrac{\pm\sqrt{1-\zeta^2}\,\omega_n}{-\zeta\omega_n} = \mp\dfrac{\sqrt{1-\zeta^2}}{\zeta}$. Time responses for different ζ $[\in [0,1)]$
and ω_n (>0) (**Figure 5.2**) show that

i)   the system is stable if the poles are all in the open left-half s plane
     (that is, the poles have negative real parts);

ii) $K_\zeta = \pm 1$ for $\zeta = 0.707$. The overshoot of a second-order system
     becomes large as $|K_\zeta|$ becomes large.

Extracting the key points from the discussion in Sections 5.1 and 5.2

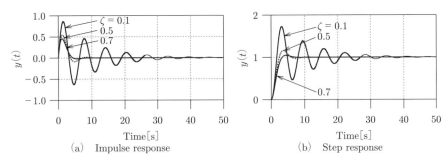

(a) Impulse response   (b) Step response

Figure 5.2 Impulse and step responses of system (5.8)

gives us the relationship between the position of poles and their impulse response (**Figure 5.3**).

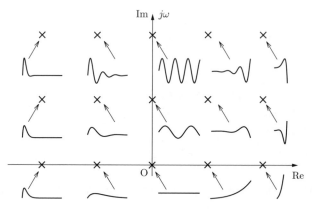

Figure 5.3 Relationship between poles and impulse responses

5.3 Routh's Stability Criterion

A typical feedback control system is shown in **Figure 5.4**. Since the feedback path does not contain any components, the system is usually referred to as a *unity-feedback system* (単位フィードバックシステム). The input-output relationship of the system is

$$G(s) = \frac{P(s)C(s)}{1 + P(s)C(s)}. \tag{5.17}$$

The poles of the closed-loop system are given by the roots of $1+P(s)C(s) = 0$.

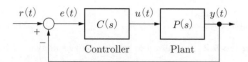

Figure 5.4 Feedback control system

The system is stable if all the poles are in the open left-half s plane and is unstable when it has any poles in the open right-half s plane. Hence, the imaginary axis, jw, is the stability boundary of a system. The behavior of a system is complex when it has poles on the imaginary axis. More specifically, the system is neutral stable[†1] if it has nonrepeated poles on the jw axis and is unstable if it has repeated poles on the jw axis.

Now, we consider a general case. Let the transfer function of a closed-loop control system be ($a_0 \neq 0$ and $m \leq n$)

$$
\begin{aligned}
G(s) = \frac{N(s)}{D(s)} &= \frac{b_0 s^m + b_1 s^{m-1} + \cdots + b_{m-1} s + b_m}{a_0 s^n + a_1 s^{n-1} + \cdots + a_{n-1} s + a_n}, \\
&= \frac{b_0 s^m + b_1 s^{m-1} + \cdots + b_{m-1} s + b_m}{(s - p_1)(s - p_2) \cdots (s - p_n)}.
\end{aligned}
\tag{5.18}
$$

The equation

$$
D(s) = a_0 s^n + a_1 s^{n-1} + \cdots + a_{n-1} s + a_n = 0
\tag{5.19}
$$

is called the *characteristic equation* (特性方程式) of the system and $D(s)$ is called the *characteristic polynomial* (特性多項式) of the system. For simplicity, we assume that $D(s) = 0$ does not have multiple roots[†2]. Since the stability of a linear system does not depend on exogenous signals, we consider the unit impulse response of the system, which is

$$
\begin{aligned}
y(t) &= \mathcal{L}^{-1} \left\{ \frac{b_0 s^m + b_1 s^{m-1} + \cdots + b_{m-1} s + b_m}{(s - p_1)(s - p_2) \cdots (s - p_n)} \right\} \\
&= \mathcal{L}^{-1} \left\{ \frac{A_1}{s - p_1} + \frac{A_2}{s - p_2} + \cdots + \frac{A_n}{s - p_n} \right\} = \sum_{i=1}^{n} A_i e^{p_i t},
\end{aligned}
\tag{5.20}
$$

where

$$
A_i = (s - p_i) \frac{N(s)}{D(s)} \Big|_{s = p_i}.
\tag{5.21}
$$

[†1] If we move the state slightly from an equilibrium but it neither tends to return to its former position nor depart from it, we say that the system is neutrally stable.

[†2] If $D(s) = 0$ has a j-multiple root, we can use the residue theorem to calculate its time response (see Section 2.3 for details).

The system asymptotically converges to zero if all the roots have negative real parts (that is, in the open left-half s plane). On the other hand, the output of the system diverges if any of the roots have positive real parts (that is, in the open right-half s plane). If we only need to know whether or not the roots of (5.19) are in the open left-half s plane, we can use some criteria to check it instead of[†] directly solving the equation.

One of the well-known methods is Routh's stability criterion. It only needs some simple calculations to obtain information about the location of the roots of $D(s)$ that determines the stability of the system. It is carried out by the following steps.

Let all the coefficients of $D(s)$ be real. without loss of generality, we assume that $a_0 > 0$. First, the following two necessary conditions must be satisfied to ensure that the poles of the system have negative real parts:

i)   None of the coefficients is zero.

ii) All coefficients of $D(s)$ have the same sign.

Next, we arrange the coefficients of $D(s)$ in the first two rows of a Routh table

1st row (s^n) a_0 a_2 a_4 $\cdots$

2nd row (s^{n-1}) a_1 a_3 a_5 $\cdots$

Then, we add subsequent rows and complete **Table 5.1**:

The elements from the third row in Table 5.1 are given as follows:

$$b_i = -\frac{1}{a_1} \begin{vmatrix} a_0 & a_{2i} \\ a_1 & a_{2i+1} \end{vmatrix} = -\frac{a_0 a_{2i+1} - a_{2i} a_1}{a_1}$$

$$\left(\text{that is, } b_1 = -\frac{a_0 a_3 - a_2 a_1}{a_1}, \ b_2 = -\frac{a_0 a_5 - a_4 a_1}{a_1}, \ \cdots \right),$$

---

†   instead of $\sim$: 〜の代わりに。

Table 5.1   Routh table

	a_0	a_2	a_4	$\cdots$	
1st row (s^n)	a_0	a_2	a_4	$\cdots$	
2nd row (s^{n-1})	a_1	a_3	a_5	$\cdots$	
3rd row (s^{n-2})	b_1	b_2	b_3	$\cdots$	
4th row (s^{n-3})	c_1	c_2	c_3	$\cdots$	
5th row (s^{n-4})	d_1	d_2	$\cdots$	$\cdots$	
$\vdots$	$\vdots$	$\vdots$	$\vdots$	$\vdots$	
n-th row (s^1)	q_1				
n-th row (s^0)	a_n				

$$c_i = -\frac{1}{b_1}\begin{vmatrix} a_1 & a_{2i+1} \\ b_1 & b_{i+1} \end{vmatrix} = -\frac{a_1 b_{i+1} - a_{2i+1} b_1}{b_1}$$

$$\left(\text{that is, } c_1 = -\frac{a_1 b_2 - a_3 b_1}{b_1}, \quad c_2 = -\frac{a_1 b_3 - a_5 b_1}{b_1}, \quad \cdots\right),$$

$$d_i = -\frac{1}{c_1}\begin{vmatrix} b_1 & b_{i+1} \\ c_1 & c_{i+1} \end{vmatrix} = -\frac{b_1 c_{i+1} - b_{i+1} c_1}{c_1}$$

$$\left(\text{that is, } d_1 = -\frac{b_1 c_2 - b_2 c_1}{c_1}, \quad d_2 = -\frac{b_1 c_3 - b_3 c_1}{c_1}, \quad \cdots\right),$$

$$\cdots,$$

where $i = 1, 2, \cdots$ Note that the elements of the third row and the subsequent rows are formed from the previous two rows.

Finally, we arrange the $(n + 1)$ elements in the first column in Table 5.1 as

$$a_0 \; a_1 \; b_1 \; c_1 \; d_1 \; \cdots \; q_1 \; a_n.$$

If they are all positive, then the roots are all in the open left-half s plane and the system is stable. However, if the signs of the elements change, then the number of the roots in the open right-half s plane is exactly the number of sign changes. For example, since the numbers of the sign changes for $+$, $-$, $+$ and $+$, $-$, $-$, $+$ are all two (one from $+$ to $-$ and the other from $-$ to $+$), two of the roots (thus, two poles of these systems) are in the open right-half s plane.

〖Example 5.1〗   Let the characteristic equation of a system be

$$D(s) = s^5 + 2s^4 + s^3 + 3s^2 + 4s + 5 = 0. \tag{5.22}$$

We use Routh's stability criterion to determine the stability of the system.

First, since all the coefficients are positive, the two necessary conditions hold. Then, we build a Routh table (**Table 5.2**) and calculate the elements:

$$b_1 = -\frac{1}{2}\begin{vmatrix} 1 & 1 \\ 2 & 3 \end{vmatrix} = -\frac{1}{2}, \ b_2 = -\frac{1}{2}\begin{vmatrix} 1 & 4 \\ 2 & 5 \end{vmatrix} = \frac{3}{2},$$

$$c_1 = -\frac{1}{-1/2}\begin{vmatrix} 2 & 3 \\ -\frac{1}{2} & \frac{3}{2} \end{vmatrix} = 9, \ c_2 = -\frac{1}{-1/2}\begin{vmatrix} 2 & 5 \\ -\frac{1}{2} & 0 \end{vmatrix} = 5,$$

$$d_1 = -\frac{1}{9}\begin{vmatrix} -\frac{1}{2} & \frac{3}{2} \\ 9 & 5 \end{vmatrix} = \frac{16}{9}.$$

Table 5.2   Routh table for (5.22)

1st row (s^5)	1	1	4
2nd row (s^4)	2	3	5
3rd row (s^3)	b_1	b_2	
4th row (s^2)	c_1	c_2	
5th row (s^1)	d_1		
6th row (s^0)	5		

Arranging the six elements in the first column yields

$$1 \quad 2 \quad -\frac{1}{2} \quad 9 \quad \frac{16}{9} \quad 5.$$

Since the number of the sign changes two times (from 2 to $-1/2$ and from $-1/2$ to 9). The system is unstable and has two poles in the open right-half s plane.

[Example 5.2]   Let $P(s) = 50/[s(s+10)]$ and $C(s) = K/(Ts+1)$ in Figure 5.4. Determine the allowable regions of K and T.

The characteristic equation of the closed-loop system is

$$1 + \frac{K}{Ts+1}\frac{50}{s(s+10)} = 0. \tag{5.23}$$

We rewrite it to be

$$Ts^3 + (10T+1)s^2 + 10s + 50K = 0. \tag{5.24}$$

When $T = 0$, (5.24) becomes

$$s^2 + 10s + 50K = 0. \tag{5.25}$$

The closed-loop system is stable if

$$K > 0. \tag{5.26}$$

When $T \neq 0$, the Routh table is shown in **Table 5.3**:

$$b_1 = -\frac{1}{10T+1}\begin{vmatrix} T & 10 \\ 10T+1 & 50K \end{vmatrix} = 10 - \frac{50KT}{10T+1}. \tag{5.27}$$

Table 5.3   Routh table for (5.24)

1st row (s^3)	T	10
2nd row (s^2)	$10T+1$	$50K$
3rd row (s^1)	b_1	
4th row (s^0)	$50K$	

Arranging the four elements in the first column yields

$$T \quad 10T+1 \quad 10 - \frac{50KT}{10T+1} \quad 50K. \tag{5.28}$$

First, all the coefficients in (5.28) must have the same sign to guarantee the stability of the system, that is,

$$T < 0, \tag{5.29}$$

$$10T + 1 < 0, \tag{5.30}$$

$$10 - \frac{50KT}{10T + 1} < 0, \tag{5.31}$$

$$50K < 0, \tag{5.32}$$

or

$$T > 0, \tag{5.33}$$

$$10T + 1 > 0, \tag{5.34}$$

$$10 - \frac{50KT}{10T + 1} > 0, \tag{5.35}$$

$$50K > 0. \tag{5.36}$$

Conditions (5.29) and (5.30) give

$$T < -0.1, \tag{5.37}$$

and Conditions (5.31) and (5.32) give

$$2 + \frac{1}{5T} < K < 0. \tag{5.38}$$

Note that (5.38) does not hold for (5.37). Thus, the system is not stable for $T < 0$.

On the other hand, Conditions (5.33) and (5.34) give

$$T > 0, \tag{5.39}$$

and Conditions (5.35) and (5.36) give

$$0 < K < 2 + \frac{1}{5T}. \tag{5.40}$$

As a result, the allowable regions of K and T are given by (5.26) for $T = 0$, and (5.39) and (5.40) for $T > 0$ (**Figure 5.5**). The following MATLAB commands draw the allowable regions:

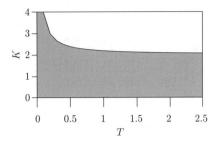

Figure 5.5   Allowable regions of K and T

```
                  Program 5.3    Allowable regions of K and T
1    T=0:0.1:2.5;
2    Tf=2.5:-0.1:0;
3
4    nT=length(T);
5    TZ=T(1,1);
6
7    TP=T(1,2:nT)';
8    KP=2+TP.^(-1)/5;
9
10   TZP=[TZ; TP; Tf'];
11   KZP=[max(KP); KP; zeros(nT,1)];
12
13   fill(TZP,KZP,'g');
14   title('Allowable regions of K and T');
15   xlabel('T'); ylabel('K');
```

5.4   Nyquist Stability Criterion

The stability of the closed-loop system in Figure 5.4 is determined by the characteristic equation of the system

$$1 + P(s)C(s) = 0. \tag{5.41}$$

Let $L(s) = P(s)C(s) = N_L(s)/D_L(s)$. Then,

$$1 + L(s) = 1 + \frac{N_L(s)}{D_L(s)} = \frac{(s - z_1)(s - z_2) \cdots (s - z_n)}{(s - p_1)(s - p_2) \cdots (s - p_n)}. \tag{5.42}$$

Since the roots of (5.41) are exactly the poles of the closed-loop system, whether or not the closed-loop system is stable [that is, all the poles of (5.17) are in the left-half s plane] is equivalent to whether or not all the roots of (5.41) have negative real parts. Thus, we can determine the system stability if we know whether or not $1 + L(s)$ has zeros in the right-half s plane or not.

Applying the *argument principle* (偏角の原理) helps us achieve the goal of determining the stability of the system. Let a contour C in the s plane encircle the entire right-half s plane, in which poles cause system unstable [**Figure 5.6**(a)]. Note that, when $L(s)$ has imaginary poles, the contour C should be modified to take a small detour around those poles to the right [Figure 5.6(b)].

$1 + L(s)$ draw a closed curve Γ on the complex plane [Figure 5.6(c)] as s travels C in the clockwise direction from the starting point 0 to $+\infty$, then to $-\infty$, and finally back to 0. When $L(s)$ has a k-tuple pole, it is called a Type-k system (see Section 6.3). Γ moves from the angle $k\pi/2$ to $-k\pi/2$ with an infinity radius in the clockwise direction when s travels C from $j0_-$ to $j0_+$ along the detour.

Let the number of unstable zeros and poles[†] of $1 + L(s)$ be Z and P, respectively. The argument principle tells us that the curve Γ encircles the origin of the $1+L(s)$ complex plane $n = P - Z$ times in the counterclockwise direction.

We need $Z = 0$ to guarantee the stability of the closed-loop system. So,

---

[†] Unstable zeros (poles) mean zeros (poles) in the right-half s plane.

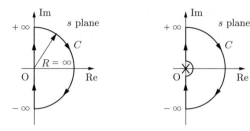

(a) Contour C encircling entier right-half s plane

(b) Modification of Contour C avoiding open-loop imaginary poles

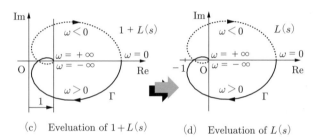

(c) Eveluation of $1+L(s)$

(d) Eveluation of $L(s)$

Figure 5.6 Nyquist plot

based on the argument principle, we know that the closed-loop system is stable if

$$n = P, \tag{5.43}$$

that is, the net number of the counterclockwise encirclements of the origin for Γ equals the number of unstable poles of $L(s)$.

Note that the difference between $1 + L(s)$ and $L(s)$ is one in the real part. Thus, if we move the imaginary axis (the vertical axis) of $1 + L(s)$ one to the right, it gives the imaginary axis for $L(s)$ [Figure 5.6(d)]. Since the origin in the complex plane for $1 + L(s)$ is the point $(-1, j0)$ in that for $L(s)$, we should counter the number of the counterclockwise encirclements of the point $(-1, j0)$ in the complex plane for $L(s)$. As a result, the *Nyquist stability criterion* (ナイキスト安定判別法) states that the closed-loop system is stable if the curve Γ encircles the point $(-1, j0)$ in the $L(s)$ complex plane P times in the counterclockwise direction. Moreover, note that the contour C in the s plane can be divided into two parts:

i)   the imaginary axis, $s = j\omega$ for $-\infty \leq \omega \leq +\infty$ and

ii)   the contour at infinity, that is, a semicircle connecting $s = +j\infty$ to $s = -j\infty$ with a radius $R = \infty$.

Since a physical system has zero response to the second part, we only need to draw the curve of $L(s)$ for the first part, which is exactly the Nyquist plot explained in Section 4.3.

A system easily becomes unstable when it has a time delay. We cannot use Routh's stability criterion to determine the stability of such a system because a time delay is not a rational function in s. However, the Nyquist stability criterion is applicable to a time-delay system.

[Example 5.3]   Check the stability of a closed-loop system with its open-loop transfer function being

$$G(s) = \frac{10}{(0.01s + 1)(0.1s + 1)(0.5s + 1)}.$$

Use the following MATLAB commands to draw a Nyquist plot for the system:

Program 5.4   Nyquist plot for $G(s)$

```
1  s=tf('s');
2  G=10/((0.01*s+1)*(0.1*s+1)*(0.5*s+1))
3  Ng=nyquistplot(G);
```

Clicking the vertical axis in the figure pops up a window for modifying the figure (**Figure 5.7**). Enlarging the vertical axis makes it easy to clearly count the number of clockwise encirclements of the point $(-1, j0)$. We can also use the command zoomcp to do so.

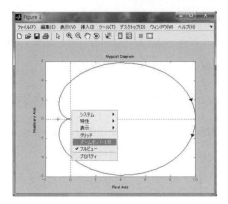

Figure 5.7 Zooming Nyquist plot

Since the number of unstable poles of $G(s)$ is $P = 0$ and the number of clockwise encirclement of the point $(-1, j0)$ is $n = 0$, the number of unstable closed-loop poles is $Z = n + P = 0$. Thus, the closed-loop system is stable.

Generally speaking, we require that a closed-loop system not only is stable but also has a certain *degree of stability* (安定度) in system design. Two quantities are commonly used to measure the stability margin for a system: *Gain and phase margins* (ゲイン余裕と位相余裕).

The Nyquist stability criterion confirms that the system is stable if the curve of $L(s)$ does not encircle the point $(-1, j0)$. It is desirable that the curve of $L(s)$ stays clear of the point $(-1, j0)$. The gain margin is a figure of merit[†] for system stability. It is defined to be

$$\text{GM} = \frac{1}{|L(j\omega_{cp})|}, \tag{5.44}$$

where $L(j\omega_{cp})$ is the point of the Nyquist plot crossing the negative real axis and ω_{cp} is called the phase crossover frequency [the phase of $L(j\omega_{cp})$ is $-180°$]. Another figure of merit for system stability is the phase margin.

---

† figure of merit：性能指数。

Let the point of the Nyquist plot crossing the unit circle be $L(j\omega_{cg})$, where ω_{cg} is called the gain crossover frequency [the gain of $L(j\omega_{cg})$ is 1]. The phase margin is defined to be

$$\text{PM} = \angle L(j\omega_{cg}) - (-180) = 180 + \angle L(j\omega_{cg}), \qquad (5.45)$$

where $\angle L(j\omega_{cg})$ is the phase lag (a number less than 0) of $L(j\omega)$ at ω_{cg}. These two margins are shown in **Figure 5.8**(a).

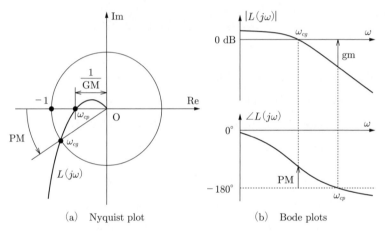

(a)   Nyquist plot   (b)   Bode plots

Figure 5.8   Definitions of gain and phase margins

5.5   Evaluation of Stability Margins using Bode Plots

The gain and phase margins, GM and PM, of the corresponding closed-loop system can be read directly from Bode plots. Note that

$$\text{gm} = 20 \log_{10} \frac{1}{\text{GM}} = -20 \log_{10} |L(j\omega_{cp})| \text{ [dB]}. \qquad (5.46)$$

For an open-loop transfer function $L(s)$, the gain margin, gm, is the vertical distance between the point $(\omega_{cp}, 20 \log_{10} |L(j\omega_{cp})|)$ and the 0-dB horizontal line in the Bode magnitude plot.

The margins for Bode plots are shown in Figure 5.8(b). A system is usually considered to be suitably designed if PM is larger or equal to[†1] $30°$. In most cases, a closed-loop system is stable if

$$\text{GM} > 1 \ (\text{gm} > 0\,\text{dB}), \ \text{PM} > 0. \tag{5.47}$$

Moreover, the larger GM (gm) and PM are, the more stable the system is. However, These are not always true. Note that the curve Γ has to encircle $(-1, j0)$ to ensure the stability of the closed-loop system when $L(s)$ has poles in the right-half s plane. So, GM < 1 is required to guarantee the stability of such a system. Another example in Advanced problem 2 in this chapter shows that these may not be true even if $L(s)$ does not have poles in the right-half s plane. We have to be aware of[†2] the difference in stability conditions for stable and unstable open-loop transfer functions and remember that (5.43) is the primary condition for system stability. It is important to keep this in mind[†3] when checking the stability of a system. It is recommended to use the Nyquist plot to check whether or not (5.47) is true before using it.

—————— **Problems** ——————

 Basic Level

[1] If the poles of a control system lie in the open right-half s plane, the system is

    A. unstable   B. stable   C. neurally stable.

[2] Consider the unity-feedback system in Figure 5.4. Find K for $C(s) = K$ when the time constants of the closed-loop systems are all set to be $0.05\,\text{s}$ for

---

[†1] be equal to ～ : ～と等しい。
[†2] be aware of ～ : ～を知っている, ～に気がついている。
[†3] keep ～ in mind : ～を心に留めておく。

    A. $P(s) = \dfrac{1}{s}$ B. $P(s) = \dfrac{50}{s+10}$.

[3] Construct a unity-feedback system for the control of a spacecraft, in which $P(s) = 1/(9s^2)$ and $C(s) = K(Ts+1)$.

    A.   Find the damping ratio and the natural angular frequency of the system based on system poles for $T = 3$ and $K = 2$.

    B.   Find the relationship between T and K if we choose the damping ratio of the control system to be $\zeta = 0.707$.

[4] Let a, b, c, d, and e be positive numbers. Which of the following systems may be stable?

    A. $as^4 + bs^3 + cs + d = 0$ B. $as^4 + bs^3 + cs^2 + ds + e = 0$

    C. $-as^4 + bs^3 + cs + d = 0$ D. $-as^4 + bs^3 + cs^2 + ds - e = 0$.

[5] Let the open-loop transfer function of a unity-feedback system be $L(s) = \dfrac{K(s+b)}{s(s+a)}$. Find the range of K that the closed-loop system is stable when

    A. $a < 0$ and $b < 0$ B. $a < 0$ and $b > 0$

    C. $a > 0$ and $b < 0$ D. $a > 0$ and $b > 0$.

[6] Use Routh's stability criterion to determine the stability of systems that has the following characteristic equations:

    A. $-s^2 - 4s - 5 = 0$ B. $s^4 + 3s^3 - 3s^2 - 7s + 6 = 0$

    C. $a_3 s^3 + a_2 s^2 + a_1 s + a_0 = 0$ D. $s^5 + 2s^4 + s^3 + 3s^2 + 4s + 5 = 0$

    E. $s^4 + 8s^3 + 10s^2 + 8s + 1 = 0$ F. $s^4 + s^3 - 3s^2 - 5s + 2 = 0$.

[7] Use Routh's stability criterion to find the range of K that ensures system stability for the following characteristic equations

    A. $s^3 + 12s^2 + 20s + 20K = 0$

    B. $s^4 + 2s^3 + Ks^2 + 10s + 100 = 0$

    C. $s^3 + 4s^2 + (K-5)s + K = 0$.

[8] Use Routh's stability criterion to find the range of $C(s) = K$ for which a unity-feedback system is stable:

    A. $P(s) = \dfrac{s^2 - 2s + 2}{s^2 + 2s + 4}$ B. $P(s) = \dfrac{0.5s + 1}{s(s+1)(0.5s^2 + s + 1)}$.

[9] A Nyquist plot for an open-loop transfer function, $L(s)$, passes through the point $(-1, \, j0)$ in $L(j\omega)$ plane. The gain margin of the corresponding closed-loop control system, GM, is

    A.   greater than one   B.   one

    C.   infinity   D.   less than one.

[10] If a system has a time delay, then we cannot directly use Routh's stability criterion to determine the stability of the system. However, we can use the Nyquist stability criterion to do it. Draw Nyquist plots for the open-loop transfer function $L(s) = e^{-Ts}/[s(s+1)(s+2)]$ for $T = 0$, 2, 4 s and determine the stability of the closed-loop systems.

[11] A Bode phase plot for an open-loop transfer function $L(s)$ at the gain cross-over frequency is $-125°$. The phase margin of the corresponding closed-loop system is

    A. $-55°$ B. $55°$ C. $-125°$ D. $125°$.

[12] Draw Nyquist plots for a unity-feedback system with each of the following open-loop transfer functions and determine the stability of the systems.

    A. $L(s) = \dfrac{1}{s^3}$ B. $L(s) = \dfrac{1}{s^2(s+1)}$

    C. $L(s) = \dfrac{50}{s(s+10)}$ D. $L(s) = \dfrac{1}{s(s+1)(2s+1)}$

    E. $L(s) = \dfrac{1}{(s+1)(2s+1)(10s+1)}$ F. $L(s) = \dfrac{10(5s+1)(12s+1)}{s(s+1)(2s+1)(10s+1)(20s+1)}$

    G. $L(s) = \dfrac{2}{s-1}$ H. $L(s) = \dfrac{1}{s(s-1)}$

    I. $L(s) = \dfrac{2(0.1s-1)}{s(0.5s+1)(s+1)}$.

[13] For the open-loop transfer functions given in Problem 12, draw their Bode plots, determine the stability of the systems, and find the gain and phase margins for stable systems.

[14] Let the open-loop transfer function be $L(s) = 1/(s^3 + 2s^2 + 2s + 1)$. Use Bode plots to find whether or not the system is stable. If it is, find the gain and phase margins. Then, use MATLAB to verify the results.

[15] Find the value of a gain, K, in a unity-feedback system for a gain margin of 10 dB for

    A. $L(s) = \dfrac{K}{s(s+4)(s+10)}$ B. $L(s) = \dfrac{K}{(s+4)(s+10)(s+15)}$.

Advanced Level

[1] Use Routh's stability criterion to determine the stability of systems that has the characteristic equation of $s^4 + 3s^3 + s^2 + 3s + 1 = 0$. Note that a coefficient in the first column in a Routh table is zero. We need to use a

small positive number ϵ to replace it to proceed with the calculation of the Routh table.

[2] GM > 1 (gm > 0 dB), PM > 0°, and $\omega_{cp} > \omega_{cg}$ hold for many stable closed-loop systems. However, these are not always true. Draw the Nyquist plot and Bode plots of $L(s) = (s+1)(2s+1)/s^3$. Verify that the closed-loop system is stable but GM < 1 (gm < 0 dB) and $\omega_{cp} < \omega_{cg}$.

[3] When an open-loop transfer function is minimum-phase and the closed-loop system is stable, the gain margin of the system, GM, is

    A. greater than one   B. infinite

    C. indefinite   D. less than one.

[4] Let $P(s) = 1/[(s+2)(s+4)(s+6)]$ in a unity-feedback system.

    A. Let the open-loop transfer function be $L(s) = KP(s)$. Use Routh's stability criterion to find the range of K that the closed-loop system is stable.

    B. Choose three values of K in the range given in A and use the Nyquist plot of $L(s)$ to check the stability of the system.

    C. Draw the Bode plots of $L(s)$ for the three values chosen in B.

    D. Use the Bode plots to verify the range of gain, K, for stability.

[5] The closed-loop transfer function of a system is

$$G(s) = \frac{s^2 + K_1 s + K_2}{s^4 + K_1 s^3 + K_2 s^2 + 4s + 1}.$$

Determine the ranges of K_1 and K_2 that guarantee the stability of the system. Use MATLAB to plot the ranges.

[6] Let $P(s) = e^{-2s}/(10s+1)$ and $C(s) = K$ in a unity-feedback system. We know that the system is stable for $0 < K < 8.56$.

    A. Can Routh's stability criterion be directly used to determine the stability of the system?

    B. Use Routh's stability criterion to find the ranges of K for the approximations of

$$e^{-2s} = 1 - 2s, \; e^{-2s} = \frac{1}{1+2s}, \text{ and } e^{-2s} = \frac{1-s}{1+s}.$$

    C. Compare the obtained ranges of K with[†] the given one.

[†] compare A with B：A を B と比較する。

【7】 Note that a stable closed-loop system does not mean that its gain and phase margins must be larger than zero. Use Routh's stability criterion to determine the stability of the following unity-feedback system and check their Nyquist and Bode plots.

A. $P(s) = \dfrac{10(s-1)}{s(10s+1)}$ and $C(s) = -0.05$.

B. $P(s) = \dfrac{7}{(s-1)(s+2)(s+4)}$ and $C(s) = 2$.

C. $P(s) = \dfrac{s-1}{s(2s+1)}$ and $C(s) = -0.5, \ -1, \ 1$.

D. $P(s) = \dfrac{4s-1}{(10s-1)(2s+1)(s+1)}$ and $C(s) = -1.5, \ -1, \ 1$.

Use the following steps to solve the above problems:

Step 1)   Use Routh's stability criterion to find the ranges of $C(s) = K$ over which the systems are stable.

Step 2)   Check whether or not the specified gains are in the ranges.

Step 3)   Draw the Nyquist plots of the systems with specified gains and use the Nyquist stability criterion to verify your answers.

Step 4)   Draw the Bode plots for the specified gains, determine whether or not the systems are stable, and find the gain and phase margins from the plots.

6

Characteristics of Control System

In this chapter, we explain the characteristics of a control system that are used to design and evaluate a controller. We first show various *transfer characteristics* (伝達特性). Then, we describe the transient and steady-state characteristics.

6.1 Transfer Characteristics

Figure 6.1 shows the block diagram of a feedback control system that contains three *exogenous signals* (外生信号): a reference input $r(t)$, a disturbance $d(t)$, and a *measurement noise* (計測騒音)[†] $n(t)$. The system has a forward path from the error to the output and a feedback path from the output back to the summer at the place of the reference input. A simple calculation gives

$$Y(s) = \frac{P(s)C(s)}{1 + P(s)C(s)}R(s) + \frac{P(s)}{1 + P(s)C(s)}D(s) - \frac{P(s)C(s)}{1 + P(s)C(s)}N(s).$$

$$(6.1)$$

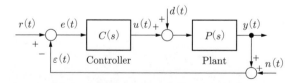

Figure 6.1   Feedback control system

[†]   While a disturbance affects the true value of a variable, a noise affects its measured value.

Let the *open-loop transfer function* (一巡伝達関数) be

$$L(s) = P(s)C(s). \tag{6.2}$$

It is the product of the transfer functions along the path from $e(t)$ to $\epsilon(t)$. (6.1) clearly shows that the stability of a linear system depends only on $1+L(s)$ but is not related to exogenous signals, including a reference input, a disturbance, and a noise.

We define the *sensitivity function* (感度関数) to be

$$S(s) = \frac{1}{1 + L(s)}, \tag{6.3}$$

and the *complementary sensitivity function* (相補感度関数) to be

$$T(s) = \frac{L(s)}{1 + L(s)}. \tag{6.4}$$

Using (6.3) and (6.4) to rewrite (6.1) yields

$$Y(s) = T(s)R(s) + P(s)S(s)D(s) - T(s)N(s). \tag{6.5}$$

Reference tracking requires that the output, $y(t)$, agrees with[†] the reference input, $r(t)$, or in other words, the error, $e(t)$, is suppressed to a low level. Since

$$E(s) = S(s)R(s), \tag{6.6}$$

good tracking performance means a small gain of $S(s)$.

It is clear from the transfer characteristics that disturbance rejection means suppressing $S(s)$ and noise reduction means suppressing $T(s)$.

The transfer function from $r(t)$ to $y(t)$ is

$$G_{yr}(s) = \frac{P(s)C(s)}{1 + P(s)C(s)}. \tag{6.7}$$

When the plant changes from $P(s)$ to $\hat{P}(s)$, it becomes

---

[†]   agree with ∼ : ∼に一致する。

$$\hat{G}_{yr}(s) = \frac{\hat{P}(s)C(s)}{1 + \hat{P}(s)C(s)}. \tag{6.8}$$

Simple manipulation yields

$$\frac{\hat{G}_{yr}(s) - G_{yr}(s)}{\hat{G}_{yr}(s)} = S(s)\frac{\hat{P}(s) - P(s)}{\hat{P}(s)}. \tag{6.9}$$

This shows how the changes in the plant influence the input-output characteristics[1]. The above relationship shows that the tracking performance of the system is guaranteed for the changes in the plant if we suppress $S(s)$.

Changes in the plant may make the control system unstable. If we describe the plant as

$$\hat{P}(s) = P(s)[1 + \Delta P_M(s)], \tag{6.10}$$

the block diagram of the control system is shown in **Figure 6.2**.

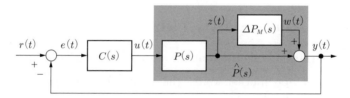

Figure 6.2 Control system contains changes in plant

Note that the stability of a linear system is independent of[2] exogenous signals. We let $r(t) = 0$ to derive a stability condition for the system in **Figure 6.3**. Defining two new signals $w(t)$ and $z(t)$ and redrawing the block diagram in Figure 6.3(a) yields Figure 6.3(b), in which

$$G_{zw}(s) = -T(s). \tag{6.11}$$

---

[1] Sensitivity is originally defined to be $S(s) = \dfrac{\Delta G_{yr}(s)/\hat{G}_{yr}(s)}{\Delta P(s)/\hat{P}(s)}$, where $\Delta G_{yr}(s) =$ $\hat{G}_{yr}(s) - G_{yr}(s)$ and $\Delta P(s) = \hat{P}(s) - P(s)$.

[2] be independent of ～ : ～と無関係で。

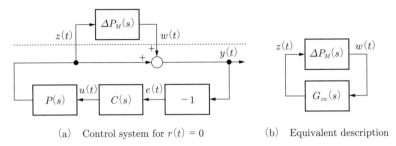

(a)   Control system for $r(t) = 0$ (b)   Equivalent description

Figure 6.3   Derivation of stability condition

Thus, $\Delta P_M(s)G_{zw}(s) = -\Delta P_M(s)T(s)$. Assuming that $\Delta P_M(s)$ is stable, we obtain the stability condition from the *small-gain theorem* (小ゲイン定理)

$$|\Delta P_M(j\omega)T(j\omega)| < 1, \ \forall \omega \in \mathbb{R}. \tag{6.12}$$

This shows that the stability of the control system can be guaranteed by suppressing $T(s)$ when the plant changes.

Since

$$S(s) + T(s) = 1, \tag{6.13}$$

we cannot suppress both $S(s)$ and $T(s)$ to a low level over the whole frequency band. Note that

i)   disturbances mainly contain low-frequency components,

ii) the steady-state characteristic (that is, a low-frequency characteristic) is a basic consideration for system design,

iii) attention is mainly paid for the changes in sensitivity in a low-frequency band for plant changes,

iv) noise mainly contains high-frequency components, and

v)   the influence of changes in a plant on the stability of a control system is mainly in a high-frequency band.

Dividing the whole frequency band into low, middle, and high is a widely used strategy that deals with a trade-off between $S(s)$ and $T(s)$. More specifically, we try to suppress the sensitivity function, $S(s)$, in the low-frequency band, to suppress the complementary sensitivity function, $T(s)$, in the high-frequency band, and to ensure a control system has enough phase and gain margins in the middle-frequency band.

【Example 6.1】   Consider the arm robot in Section 3.1 (Figure 3.3). The transfer function of the arm robot is

$$P(s) = \frac{\beta}{s(s+\alpha)}. \tag{6.14}$$

Let the nominal values of the parameters be

$$\alpha_0 = 6.25, \ \beta_0 = 33.2. \tag{6.15}$$

If we use a PD controller $C(s) = 0.294s + 3.01$, then we have

$$L(s) = \frac{33.2(3.01s + 0.294)}{s^2 + 6.25s},$$
$$S(s) = \frac{9.76s + 99.9}{s^2 + 16.0s + 99.9},$$
$$T(s) = \frac{s^2 + 6.25s}{s^2 + 16.0s + 99.9}. \tag{6.16}$$

Bode plots for the control system are obtained by the following MAT-LAB commands:

```
Program 6.1   S(s) & T(s)

1  s=tf('s'); % Laplace operator
2  alpha0=6.25; beta0=33.2; % Plant parameters
3  P=beta0/(s^2+alpha0*s) % Plant
4  L=P*(0.294*s+3.01) % Open-loop transfer function
5  S=L/(1+L), S=minreal(S), % Sensitivity function
6  T=1-S % Complementary sensitivity function
7  bode(S,T); % Bode plots
8  allmargin(L)
```

It yields PM $= 76.8°$. Since the phase of $L(j\omega)$ does not cross $-180°$, only PM is displayed. The plots (**Figure 6.4**) show that the gain of the sensitivity function is low at low frequencies (it is lower than $-20\,\mathrm{dB}$ for $\omega < 1\,\mathrm{rad/s}$) and it becomes larger as the frequency becomes larger. The complementary sensitivity function is low at high frequencies (it is lower than $-20\,\mathrm{dB}$ for $\omega > 100\,\mathrm{rad/s}$) and it becomes larger as the frequency becomes lower.

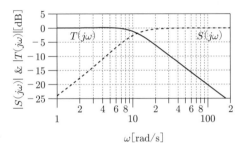

Figure 6.4 $S(s)$ & $T(s)$

When the parameters change to

$$\alpha = 4.17, \ \beta = 22.1, \tag{6.17}$$

a simple calculation gives

$$\Delta P_M(s) = \frac{22.1}{s^2 + 4.17s} \bigg/ \frac{33.2}{s^2 + 6.25s} - 1 = -\frac{11.1s + 0.319}{33.2(s + 4.17)}. \tag{6.18}$$

$\max |\Delta P_M(j\omega)T(j\omega)|$ is calculated using the following commands:

```
1  G=dPM*T
2  norm(G,Inf) % H-inf norm
```

The norm is 0.3343. Thus, the system is robustly stable with respect to[†] the parameter changes [from (6.15) to (6.17)].

[†]   with respect to ~ : ～に関して。

6.2   Transient Characteristics

When a feedback control system is stable, the response for a reference input gradually converges at a steady state in which behavior is the same over repeated observations. Such a response is called a *steady-state response* (定常応答). Before a system enters the steady state, it is in a transient state, and the corresponding response is called a *transient response* (過渡応答).

A unit step function is widely used as a reference input in a servo system or a process control system. Taking a step response as an example, we consider performance specifications in the time domain. Let the output in the steady state be $y(\infty)$. Parameters that characterize a control system are as follows[†] (**Figure 6.5**):

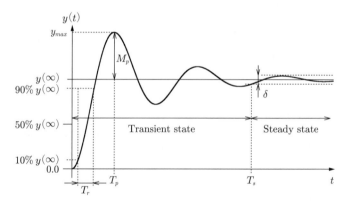

Figure 6.5   Step response

(1) Rise time T_r   the time in the response from $10\%y(\infty)$ to $90\%y(\infty)$.

(2) Overshoot M_p   a parameter related to the maximum amount

---

[†]   Except for the above-mentioned three parameters, the delay time, the peak time, the decay ratio, and others are also used to characterize a step response. However, the above three parameters are enough to describe the response.

of the response $(y_{\max})$ that is defined to be

$$M_p\% = \frac{y_{\max} - y(\infty)}{y(\infty)} \times 100\%. \tag{6.19}$$

(3) Settling time T_s the minimum time after which the response remains within a specified tolerance band δ of the steady-state value. δ is usually chosen to be 5%, 2%, or 1%.

These three parameters are used as standard time-domain specifications for a step response.

As explained in Section 5.2, the step response of a standard second-order system

$$G(s) = \frac{\omega_n^2}{s^2 + 2\zeta\omega_n s + \omega_n^2}, \tag{6.20}$$

for $0 \leq \zeta < 1$ and $\omega_n > 0$ is

$$\begin{cases} y(t) = 1 - \dfrac{1}{\sqrt{1-\zeta^2}} e^{-\zeta\omega_n t} \sin(\omega_d t + \phi), \\[2mm] \omega_d = \sqrt{1-\zeta^2}\,\omega_n, \quad \phi = \tan^{-1} \dfrac{\sqrt{1-\zeta^2}}{\zeta}. \end{cases} \tag{6.21}$$

Since $\mathrm{d}y(t)/\mathrm{d}t = 0$ when $y(t)$ reaches its maximums, we calculate the derivative of $y(t)$ and obtain

$$\omega_d T_p = \pi \tag{6.22}$$

for the overshoot, which is the first maximum. Substituting (6.22) into (6.21) and compute M_p from $y(T_p) = 1 + M_p$, we have

$$M_p = e^{-\pi\zeta/\sqrt{1-\zeta^2}}, \quad 0 \leq \zeta < 1. \tag{6.23}$$

The relationship between M_p and ζ is shown in **Figure 6.6** and **Table 6.1**.

A MATLAB command `fit` enables us to find a third-order approximate relationship

$$\zeta \doteq -2.412 M_p^3 + 3.853 M_p^2 - 2.518 M_p + 0.820, \tag{6.24}$$

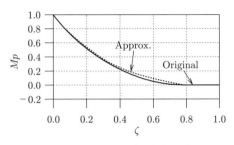

Figure 6.6 M_p vs. ζ, (6.23)

Table 6.1   Relationship between M_p and ζ

ζ	0.1	0.2	0.3	0.4	0.5	0.6	0.7	0.8
M_p	0.729	0.527	0.372	0.254	0.163	0.095	0.046	0.015

which can be used to easily find a ζ for a given M_p in system design.

We cannot find a precise analytical relationship between T_r, ω_n, and ζ. However, if we let $\tau = \omega_n t$, we can use a MATLAB command `stepinfo` to easily find $\omega_n T_r$ for different ζ [Note that $\omega_n T_r$ is exactly the rise time of the system $1/(s^2 + 2\zeta s + 1)$] (**Table 6.2**).

Table 6.2   Relationship between ζ and $\omega_n T_r$

ζ	0.1	0.2	0.3	0.4	0.5	0.6	0.7	0.8	0.9
$\omega_n T_r$	1.127	1.206	1.324	1.465	1.639	1.856	2.127	2.468	2.883

Using a MATLAB command `fit` for the date in Table 6.2 yields

$$\omega_n T_r \doteq 1.513\zeta^3 + 0.022\zeta^2 + 0.799\zeta + 1.041. \tag{6.25}$$

Note that the deviation of $y(t)$ from its steady-state value $y(\infty)$ in a step response (6.21) is the product of the exponential decay function and a sine function. It is precise enough to use the exponential envelope curve

$$y_e(t) = 1 \pm \frac{1}{\sqrt{1-\zeta^2}}e^{-\zeta\omega_n t} \tag{6.26}$$

to calculate the settling time (**Figure 6.7**). Thus, we have

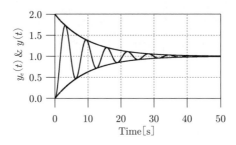

Figure 6.7   Exponential envelope $y_e(t)$, (6.26)

$$\frac{e^{-\zeta\omega_n T_s}}{\sqrt{1-\zeta^2}} \doteqdot \delta,\tag{6.27}$$

that is,

$$\omega_n T_s \doteqdot \frac{-\ln\delta - \ln\sqrt{1-\zeta^2}}{\zeta}.\tag{6.28}$$

Since $\delta \ll \sqrt{1-\zeta^2}$ for $\zeta < 0.8$,

$$\omega_n T_s \doteqdot \frac{-\ln\delta}{\zeta}.\tag{6.29}$$

The relationship between $\omega_n T_s$ and ζ, (6.29), is shown in **Figure 6.8**[†].

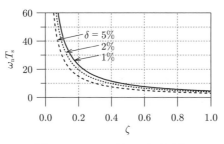

Figure 6.8 $\omega_n T_s$ vs. ζ, (6.29)

On the other hand, the open-loop transfer function of (6.20) is

$$L(s) = \frac{\omega_n^2}{s(s + 2\zeta\omega_n)}.\tag{6.30}$$

[†] $-\ln 0.01 = 4.6$, $-\ln 0.02 = 4$, and $-\ln 0.05 = 3$.

From

$$|L(j\omega_{cg})| = 1 \quad \Rightarrow \quad \omega_{cg}^4 + 4\zeta^2\omega_n^2\omega_{cg}^2 = \omega_n^4, \tag{6.31}$$

we have

$$\frac{\omega_n^2}{\omega_{cg}\sqrt{\omega_{cg}^2 + (2\zeta\omega_n)^2}} = 1.$$

Thus,

$$\omega_{cg} = \sqrt{\sqrt{4\zeta^4 + 1} - 2\zeta^2}\,\omega_n. \tag{6.32}$$

The PM is

$$\mathrm{PM} = 180° + \left(-90° - \tan^{-1}\frac{\omega_{cg}}{2\zeta\omega_n}\right) = \tan^{-1}\frac{2\zeta\omega_n}{\omega_{cg}}$$

$$= \tan^{-1}\frac{2\zeta}{\sqrt{\sqrt{4\zeta^4 + 1} - 2\zeta^2}}. \tag{6.33}$$

The relationship between PM and ζ is shown in **Figure 6.9**. Note that

$$\mathrm{PM} = 100\zeta \tag{6.34}$$

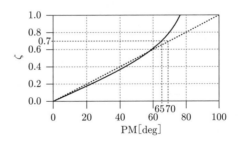

Figure 6.9 ζ vs. PM

is a good approximation for $0 < \zeta \leqq 0.7$. This relationship provides us a way to choose a ζ for a given PM and vice versa.

Moreover, the solution of $\mathrm{d}|G(j\omega)|/\mathrm{d}\hat{\omega} = 0$ ($\hat{\omega} = \omega/\omega_n$) gives the maximum of $|G(j\omega)|$:

$$|G_{\max}(j\omega)| = \frac{1}{2\zeta\sqrt{1-\zeta^2}}, \ \frac{\omega}{\omega_n} = \sqrt{1-2\zeta^2}. \tag{6.35}$$

Clearly, if we choose

$$\zeta_{opt} = \frac{1}{\sqrt{2}} = 0.707, \tag{6.36}$$

then the system does not show a peak above zero frequency. ζ_{opt} is widely used in control system design.

The above observations reveal the follows:

i)   For a fixed ω_n, we need to decrease ζ if we want to shorten T_r. On the other hand, we need to increase ζ if we want to shorten T_s.

ii) Increasing ω_n shortens all T_r, T_p, and T_s.

iii) The overshoot M_p is solely determined by ζ. The smaller ζ is, the bigger M_p will be.

6.3   Steady-State Characteristics

After a transient state, a system enters and stays in a steady state (Figure 6.5). There is a *steady-state error* (定常偏差) between a reference input and an output. It is one of the key *performance indexes* (評価指標) that evaluate the effectiveness of a feedback control action.

Let $d(t) = 0$ and $n(t) = 0$ in Figure 6.1. The steady-state error for $r(t)$ is

$$E(s) = \frac{1}{1+L(s)}R(s) = S(s)R(s). \tag{6.37}$$

If the steady-state error is a constant, we can use the final-value theorem in Table 2.2 to calculate it. The theorem is restated below:

Final-value theorem   Assume that the Laplace transform of $f(t)$ is $F(s)$. If $f(\infty)$ exists, then

$$f(\infty) = f(t)|_{t=\infty} = \lim_{s \to 0} sF(s). \tag{6.38}$$

Let the open-loop transfer function be

$$L(s) = \frac{K \prod_{j=1}^{m}(s + z_j)}{s^k \prod_{i=1}^{n}(s + p_i)} = \frac{K(s + z_1)(s + z_2)\cdots(s + z_m)}{s^k(s + p_1)(s + p_2)\cdots(s + p_n)}, \tag{6.39}$$

where K is the gain, nonzero p_i $(i = 1, 2, \cdots, n)$ are the open-loop poles, and nonzero z_j $(j = 1, 2, \cdots, m)$ are the open-loop zeros. We call the control system a Type-k system because it has a k-tuple pole at the origin. The steady-state errors for different reference inputs can easily be calculated using (6.37) and (6.38), for example, the steady-state error of a Type-0 system, $e_{ss} = e(\infty)$, for a unit step reference input is

$$e_{ss} = \lim_{s \to 0} s \frac{1}{1 + \frac{K \prod_{j=1}^{m}(s + z_j)}{\prod_{i=1}^{n}(s + p_i)}} \frac{1}{s} = \frac{1}{1 + K_p}, \quad K_p = \lim_{s \to 0} L(s); \tag{6.40}$$

that of a Type-1 system for a unit ramp reference input is

$$e_{ss} = \lim_{s \to 0} s \frac{1}{1 + \frac{K \prod_{j=1}^{m}(s + z_j)}{s \prod_{i=1}^{n}(s + p_i)}} \frac{1}{s^2} = \frac{1}{K_v}, \quad K_v = \lim_{s \to 0} sL(s); \tag{6.41}$$

and that of a Type-2 system for a unit parabola reference input is

$$e_{ss} = \lim_{s \to 0} s \frac{1}{1 + \frac{K \prod_{j=1}^{m}(s + z_j)}{s^2 \prod_{i=1}^{n}(s + p_i)}} \frac{1}{s^3} = \frac{1}{K_a}, \quad K_a = \lim_{s \to 0} s^2 L(s). \tag{6.42}$$

Some typical results are shown in **Table 6.3** and **Figure 6.10**.

Table 6.3 Relationship between steady-state errors and system types

System type	Step $[r(t) = 1(t)]$	Ramp $[r(t) = t]$	Parabola $\left[r(t) = \dfrac{1}{2}t^2\right]$
0	$\dfrac{1}{1+K_p}$	∞	∞
1	0	$\dfrac{1}{K_v}$	∞
2	0	0	$\dfrac{1}{K_a}$
3	0	0	0

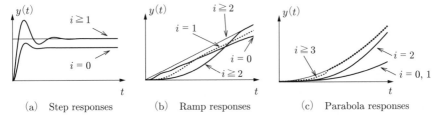

(a) Step responses   (b) Ramp responses   (c) Parabola responses

Figure 6.10 Type i vs. steady-state error

Problems

⟨ **Basic Level** ⟩

[1] Let $P(s) = 50/[s(s+10)]$ and $C(s) = 50(1+1/(0.1s)+0.032s)$ in Figure 6.1. Find the transfer function $G_{yd}(s)$ and draw Bode plots for it. Use the Bode magnitude plot to explain how the system suppresses a disturbance $d(t) = \sin t$ and use the following MATLAB commands to verify it:

`Gyd=P/(1+P*C), bode(Gyd), t=0:0.01:20; d=sin(t); lsim(Gyd,d,t);`

[2] Let $P(s) = 50/[s(s+10)]$ and $C(s) = 50(1+1/(0.1s)+0.032s)$ in Figure 6.2. When the plant contains a time delay, that is, $\hat{P}(s) = 50e^{-Ls}/[s(s + 10)]$. Use the following steps to find out how large the time delay does not destroy the stability of the system for the PID controller:

Step 1) Find the complementary sensitivity function $T(s) = P(s)C(s)/[1 + P(s)C(s)]$.

Step 2) Verify $\Delta P_M(s) = \hat{P}(s)/P(s) - 1 = e^{-Ls} - 1$.

Step 3) Let $W(s) = 2.1Ls/(Ls + 1)$. Use the Bode plots to verify $|e^{-jL\omega} - 1| \leq |W(j\omega)|$.

Step 4) Since the closed-loop stability is guaranteed by $|\Delta P_M(j\omega)T(j\omega)| \leq |W(j\omega)T(j\omega)| < 1$ [the small-gain theorem, (6.12)], draw the Bode magnitude plot of $W(s)T(s)$ for different L and check if the maximum gain is less than 1 to find the largest value of L.

[3] Find ζ and ω_n of a second-order system $G(s) = 36/(s^2 + 6s + 36)$.

[4] A system has a single pole at the origin. The impulse response of the system

    A. is a constant   B. decays exponentially

    C. is a ramp   D. oscillates.

[5] For a Type-1 system, the steady-state error, e_{ss}, for a step input is

    A. ∞ B. 0   C. 0.25   D. 0.5.

[6] For a Type-2 system, e_{ss} for a ramp input is

    A. 0   B. ∞ C. a non-zero number.

[7] Use MATLAB to find the unit ramp response of the transfer function $G(s) = 20/(s^2 + 4s + 25)$.

[8] Consider a unity-feedback system in which $P(s) = 1/[s(0.1s + 1)]$ and $C(s) = 10$.

    A. Determine the damping ratio and natural angular frequency of the closed-loop system.

    B. Find the rise time, overshoot, and settling time ($\delta = 5\%^{\dagger}$) of the unit step response for the system.

    C. Calculate the steady-state errors of the control system for the inputs of a step, ramp, and parabolic signals.

[9] The forward transfer function of a unity-feedback system is $L(s) = 1\,000(s + 8)/[(s + 7)(s + 9)]$

    A. Find the type of the system.

    B. Find the steady-state errors for the step, ramp, and parabolic inputs.

$\dagger$ The default value of δ is set to be 2% for calculating a settling time in MATLAB.

Advanced Level

[1] Consider the control system in Figure 6.1, in which the plant is $P(s) = \beta/[s(s+\alpha)]$. Let the nominal values of the parameters be $\alpha_0 = 6.25$ and $\beta_0 = 33.2$; and the changed values, $\alpha = 4.17$ and $\beta = 22.1$. For a PD controller $C(s) = 0.323s + 4.337$, draw the Bode plots of the open-loop transfer function, the sensitivity function, and the complementary sensitive function of the control system for the nominal and the changed values. Find the changes in gm and PM caused by the changes in α and β. Explain the disturbance-suppression characteristics of the system for $d(t) = \sin t$ and $d(t) = \sin 5t$ based on the Bode plots of the sensitive functions.

[2] For a standard second-order system $G_{yr}(s) = \omega_n^2/(s^2 + 2\zeta\omega_n s + \omega_n^2)$, the step response is $y(t) = 1 - \dfrac{1}{\sqrt{1-\zeta^2}} e^{-\zeta\omega_n t} \sin(\omega_d t + \phi)$ for $0 \le \zeta < 1$, where $\omega_d = \sqrt{1-\zeta^2}\,\omega_n$ and $\phi = \tan^{-1} \dfrac{\sqrt{1-\zeta^2}}{\zeta}$. The trancking error is $e(t) = \dfrac{1}{\sqrt{1-\zeta^2}} e^{-\zeta\omega_n t} \sin(\omega_d t + \phi)$. Find the integral absolute error $\displaystyle\int_0^\infty |e(t)|\,\mathrm{d}t$.

[3] If the steady state of a system is not a constant value, we cannot use the final-value theorem to find the steady-state error. However, we can use the frequency response of the system to calculate it. Let $P(s) = 1/(s+1)$ and $C(s) = 1$ in a unity-feedback system. Find the steady-state error for $r(t) = \sin(t + 30°) - \cos(2t - 45°)$.

[4] Let $P(s) = K_2/[s(s+1)(s+2)]$ and $C(s) = K_1(s+3)/s$ in Figure 6.1.
   A. Calculate e_{ss} for $r(t) = 1(t) + 2t + 0.3t^2$.
   B. Calculate e_{ss} for $d(t) = \delta(t) + 1(t)$.

[5] Consider a control system in **Figure 6.11**.
   A. For $K_F = 0$, determine K_P that yields $e_{ss} = 0.05$ for $d(t) = 1(t)$.

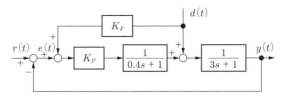

Figure 6.11 Advanced problem 5

B.   For the selected K_P, determing K_F that ensures $e_{ss} = 0$ for $d(t) = 1(t)$.

[6]   Consider a control system in **Figure 6.12**, in which $P(s) = 50/s(s+10)$ and $C_F(s) = (b_2 s^2 + b_1 s)/(Ts + 1)$.

A.   Find the steady-state tracking error for $r(t) = (1/2)t^2$ when $C_F(s)$ is not used.

B.   Find the parameters b_1 and b_2 ensuring that the steady-state tracking error is zero by inserting $C_F(s)$.

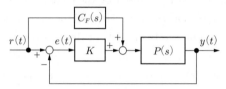

Figure 6.12   Advanced problem 6

7 Transfer-Function-based Control-System Design

In this chapter, we consider the problem of system design in the frequency domain based on the techniques explained in previous chapters. The internal-model principle, the degree of freedom in a control system, and other important concepts are presented in this chapter to help with designing a high-performance control system.

7.1   Basic Consideration in Control-System Design

We first consider a control system in the frequency domain. As explained in the Nyquist stability criterion, the relationship between the Nyquist plot of an open-loop transfer function $L(s)$ and the point $(-1, j0)$ in the s plane can be used to determine the stability of a control system. However, only ensuring the stability of a system is not enough for control-system design because changes in a plant may cause a control system unstable. For this reason, we usually design a control system that has adequate stability margins.

As explained in Sections 5.4 and 5.5, for a minimum-phase open-loop $L(s)$, the gain margin of a system (gm) is the difference between the open-loop gain $|L(j\omega_{cp})|$ and $0\,\mathrm{dB}$, and the phase margin (PM) is the difference between the phase $\angle G(j\omega_{cg})$ and $-180°$ [Figure 5.8].

Generally speaking, a desirable control system usually has gm of $10 \sim$

$20\,\mathrm{dB}$ and PM of $30° \sim 60°$. PM $= 30°$ is usually used as a lower bound in control system design. ω_{cg} is chosen to be large enough such that the closed-loop bandwidth satisfies prescribed control specifications.

Now, we examine some control systems in the time domain.

There are two parameters, T and K, in a first-order system $G(s) = K/(Ts + 1)$. The pole of the system is $p = -1/T$. The time constant, T, determines the transient response of the system; and the gain, K, the steady-state characteristic. The step response is

$$y(t) = K\left(1 - e^{-t/T}\right). \tag{7.1}$$

A simple calculation shows that the rise time and the settling time of the system are[†]

$$T_r = 2.2T, \quad T_s = 3T \text{ (for } \delta = 5\%). \tag{7.2}$$

The following results are useful for system design:

i)   The pole should be chosen in the open left-half s plane to guarantee the stability of the system.

ii) The farther the pole is from the origin, the faster the system response will be.

There are two parameters, ζ and ω_n, in a standard second-order system

$$G(s) = \frac{\omega_n^2}{s^2 + 2\zeta\omega_n s + \omega_n^2}. \tag{7.3}$$

When we talk about a second-order system, we usually refer to a system that has not only an order of two but also an oscillating characteristic, that is, $0 \leqq \zeta < 1$, or in other words, it has a pair of complex conjugate poles $p_{1,2} = -\zeta\omega_n \pm j\sqrt{1 - \zeta^2}\omega_n$. The gradients of the poles are $K_\zeta = \pm\sqrt{1 - \zeta^2}\omega_n/(-\zeta\omega_n) = \mp\sqrt{1 - \zeta^2}/\zeta$.

[†]   The settling time is $4.6T$ for $\delta = 1\%$ and $4T$ for $\delta = 2\%$ for a first-order system.

Since the relationship between the rise time and the parameters ζ and ω_n is nonlinear [see (6.25)], we consider the step response curve for ζ around 0.6 on an average[†]. It is

$$T_r \doteqdot \frac{1.8}{\omega_n}. \tag{7.4}$$

The overshoot is [see (6.23)]

$$M_p = e^{-\frac{\pi\zeta}{\sqrt{1-\zeta^2}}}, \quad 0 \leq \zeta < 1, \tag{7.5}$$

and its approximate relationship is given in (6.24). The settling time is

$$T_s = \frac{4.6}{\sigma}, \quad \sigma = \zeta\omega_n \tag{7.6}$$

for $\delta = 1\%$ [see (6.29)].

The *pole-placement method* (極配置法) is a widely used strategy for system design. A general guideline for the design of a step response is that we need to choose a small T_r for a quick response, small M_p and T_s for a good damping characteristic, and a small $|e(\infty)|$ for a good tracking characteristic.

Below, we show how to specify a region in the s plane for prescribed time-domain specifications T_r, M_p, and T_s to perform pole placement for a second-order system.

First, we obtain

$$\omega_1 = \frac{1.8}{T_r}, \quad K_1 = \mp\frac{\sqrt{1-\zeta_1^2}}{\zeta_1}, \quad \sigma_1 = \frac{4.6}{T_s} \tag{7.7}$$

from (7.4), (6.24), and (7.6). Note that, if the poles of the designed system are in the shadow regions (a), (b), and (c) in **Figure 7.1**, then the rise time, the overshoot, and the settling time are less or equal to T_r, M_p,

---

[†] on an average : 平均して。

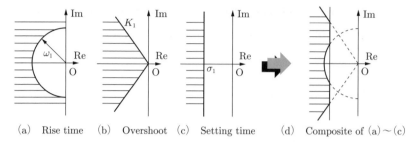

(a)   Rise time   (b)   Overshoot   (c)   Setting time   (d)   Composite of (a)∼(c)

Figure 7.1   Region in s plane for pole placement
characterized by control specifications

and T_s, respectively. So, if we choose the poles in the shadow region in
Figure 7.1(d), which is the region of the composite of (a)∼(c) in Figure 7.1,
the designed system satisfies all the three time-domain specifications.

Note that the above discussions are correct only for the standard second-
order system (7.3). A complicated system has more poles and may even
have zeros. Even though the relationships (7.4), (6.24), and (7.6) are not
precisely held for a complicated system, they can be used as guidelines for
system design.

Investigate the following two cases helps with understanding the effects
of additional zeros and poles on the system (7.3)

$$G_z(s) = \frac{\dfrac{s}{\alpha \zeta \omega_n} + 1}{\left(\dfrac{s}{\omega_n}\right)^2 + 2\zeta \dfrac{s}{\omega_n} + 1}, \tag{7.8}$$

and

$$G_p(s) = \frac{1}{\left[\dfrac{s}{\alpha \zeta \omega_n} + 1\right]\left[\left(\dfrac{s}{\omega_n}\right)^2 + 2\zeta \dfrac{s}{\omega_n} + 1\right]}, \tag{7.9}$$

where α is a constant.

When the system (7.3) has an additional zero $z = -\alpha \zeta \omega_n = -\alpha \sigma$, the
zero has little effect on the system response as α is large because the zero

is far from the poles in the left-half s plane. When $\alpha = 1$, the zero has the same real part as the poles have and gives a large effect on the system response. The observation of the step response for different α [**Figure 7.2**(a)] shows that the overshoot of the transient response of the system becomes larger as the zero becomes closer to the imaginary axis. But the zero gives little influence on the settling time.

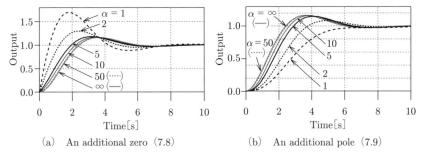

(a)   An additional zero (7.8)   (b)   An additional pole (7.9)

Figure 7.2   Effects of additional zero or pole for $\zeta = 0.5$ and $\omega_n = 1$

When the system (7.3) has an additional pole $p = -\alpha\zeta\omega_n = -\alpha\sigma$, the pole mainly gives influence on the rise time [Figure 7.2(b)]. The rise time becomes longer as the pole becomes closer to the imaginary axis.

As guidelines for system design, we should increase ω_n if we want to shorten T_r, we should increase ζ if we want to suppress M_p, and we should move poles to the left if we want to reduce T_s. Figure 7.2 shows that additional zeros and/or poles give a small effect on system response if $\alpha \geqq 5$. This observation suggests a method of choosing poles of a system, the dominant-pole approximation. We divide the poles of the system into two groups: one is close to the imaginary axis. Poles in this group, which are called *dominant poles* (代表極), dominate system response. The other one is a group in which the real parts of the poles are about $5 \sim 10$ times the ones of the dominant poles. Poles in this group give little effect on system response. The system behaves like a first-order system if the system only has one dominant pole, and like a second-order system if it has two conjugate dominant poles.

7.2   Internal-Model Principle

Consider the feedback control system in **Figure 7.3**. Let the Laplace transform of a reference input and a disturbance be

$$R(s) = \frac{N_r(s)}{D_r(s)}, \quad D(s) = \frac{N_d(s)}{D_d(s)}. \tag{7.10}$$

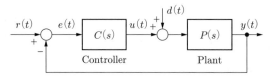

Figure 7.3   Feedback control system

Assume that all the poles in $D_r(s)$ and $D_d(s)$ are unstable and the plant does not contain $D_r(s)$ or $D_d(s)$ as its zeros. Let the plant and the controller be

$$P(s) = \frac{N_p(s)}{D_p(s)}, \quad C(s) = \frac{N_c(s)}{D_c(s)}. \tag{7.11}$$

First, consider the problem of tracking the reference input. A simple calculation shows that the *tracking error* (追従誤差)[†] $E(s) = R(s) - Y(s)$ is

$$E(s) = \frac{1}{1 + P(s)C(s)}R(s) = \frac{D_p(s)D_c(s)}{D_p(s)D_c(s) + N_p(s)N_c(s)}\frac{N_r(s)}{D_r(s)}. \tag{7.12}$$

We aim to design a controller that ensures $e_{ss} = \lim_{t \to \infty} e(t) = 0$.

The internal-model principle states that the output tracks the reference input without steady-state tracking error if and only if

---

[†]   An error is also called a tracking error in a servo system.

i)   the closed-loop system is stable and

ii) $D_r(s)$ is contained in $D_p(s)D_c(s)$.

Since $D_r(s)$ is usually not contained in $D_p(s)$, we need to design a controller in which $D_c(s)$ contains $D_r(s)$.

The problem of rejecting an exogenous disturbance can be considered in the same manner[†]. It is easy to show that we require that $D_d(s)$ is contained in $D_p(s)D_c(s)$ if we want to completely reject the disturbance in the steady state, that is, we usually need to insert $D_d(s)$ in $D_c(s)$.

The internal-model principle plays a crucial role in servo systems.

7.3   Design of PID-Control System

Proportional-integral-derivative (*PID*) *control* (PID 制御) is widely used in many control systems. Statistics show that more than 90% of control loops use PID control in process industries. This control method has many characteristics: It balances the information in the past, at present, and in the future in a control action; it has a simple structure and only has three tuning parameters; and it is easy to implement.

The PID controller in a PID-control system (**Figure 7.4**) is

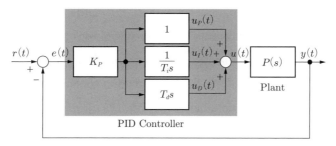

Figure 7.4   PID-control system

$$C(s) = K_P \left(1 + \frac{1}{T_i s} + T_d s \right), \tag{7.13}$$

where K_P is the proportional gain, T_i is the integration time, and T_d is the derivative time.

The proportional control term produces a control input

$$U_P(s) = K_P E(s), \quad u_P(t) = K_P e(t). \tag{7.14}$$

It yields a control action in proportion to the error, which is the most standard control action.

The integral control term produces a control input

$$U_I(s) = \frac{K_P}{T_i s} E(s), \quad u_I(t) = \frac{K_P}{T_i} \int_0^t e(\tau) d\tau. \tag{7.15}$$

It yields a control action based on the accumulated error. This control law suppresses the error in the steady state.

The derivative control term produces a control input

$$U_D(s) = K_P T_d s E(s), \quad u_D(t) = K_P T_d \frac{de(t)}{dt}. \tag{7.16}$$

Note that

$$\frac{de(t)}{dt} \doteqdot \frac{e(t+T) - e(t)}{T} \tag{7.17}$$

holds for a small positive number T and $e(t+T)$ is the error in T seconds in the future. While proportional-integral control uses the error at present and in the past, derivative control takes account of[†] the error in the future. This control law estimates the changing trend in the error before it starts to change. This makes it possible to suppress the error in a timely fashion.

A *frequency band* (周波数帯域) is a frequency range in which input signals pass through a system without attenuation. As shown in **Figure 7.5**, the

---

†   take account of 〜：〜を考慮に入れる，斟酌する。

cutoff angular frequency (遮断角周
波数) of a first-order system is ω_c
(that is, the system gain decreases
to $-3\,\mathrm{dB}$ or the amplitude of the
output of a system is reduced to
0.707 times of the input at ω_c). Thus, the angular frequency band of
the system is $[0, \omega_c]$.

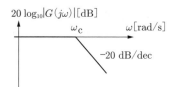

Figure 7.5 Definition of frequency band

The suitable use of the derivative control can improve the phase margin
of a control system. So, it reduces the overshoot of the system and enhances
the degree of stability. It can also enlarge the cutoff angular frequency of a
closed-loop system. Thus, it results in quicking the response of the system.

The integral control mainly affects the steady state; and the derivative
control, the transient response. As a matter of course[†1], we do not take the
whole PID actions as the first choice in the design of a control law. Note
that integral control has a high gain at a low frequency; and derivative
control, at a high frequency. We choose PI control if we focus on steady-
state control precision, and PD control if we focus on transient control
performance. However, if none of these two combinations can provide us
with[†2] satisfactory control results, then we start to consider using the whole
three PID actions.

A large number of PID design methods have been devised. A step response is usually used to determine the parameters. We explain two methods in this section.

(1) Ziegler-Nichols tuning methods   This method is the most
widely used one. For a plant

[†1]   as a matter of course：もちろん（のこと），当然。
[†2]   provide A with B：A に B を供給する，与える。

$$P(s) = \frac{Ke^{-sL}}{Ts+1} \tag{7.18}$$

or

$$P(s) = \frac{Ke^{-sL}}{Ts}. \tag{7.19}$$

the parameters obtained by the Ziegler-Nichols tuning method yield the decay ratio of one-quarter[†] for a step response (**Table 7.1**). Note that the parameters are chosen such that the proportional control yields the decay ratio of one-quarter, and the other control yields almost the same decay ratio.

Table 7.1   Ziegler-Nichols tuning method for decay ratio of one-quarter

Controller	K_P	T_i	T_d
P	$\dfrac{T}{LK}$	–	–
PI	$\dfrac{0.9T}{LK}$	$3.3L$	–
PID	$\dfrac{1.2T}{LK}$	$2.0L$	$0.5L$

The *ultimate sensitivity method* (限界感度法) is an approach for a plant having a general form. First, consider the step response of a proportional-control system for a plant. Increase K_P to the ultimate gain K_c, at which the output has continuous oscillations with a constant amplitude. Let the oscillation period (called the ultimate period) be T_c. Then, the Ziegler-Nichols tuning method uses K_c and T_c to give the PID parameters (**Table 7.2**).

Note that

$$T_i = 4T_d \tag{7.20}$$

was recommended both in Tables 7.1 and 7.2.

[†]   Let the i-th overshoot be M_{pi} and the $(i+1)$-th be $M_{p(i+1)}$. The decay ratio of one-quarter means that $M_{p(i+1)}/M_{pi} = 1/4$.

Table 7.2 Ziegler-Nichols tuning method
based on ultimate sensitivity method

Controller	K_P	T_i	T_d
P	$0.5K_c$	–	–
PI	$0.45K_c$	$0.83T_c$	–
PD	$0.8K_c$	–	$0.125T_c$
PID	$0.6K_c$	$0.5T_c$	$0.125T_c$

(2) Partial-model-matching method   The input-output transfer
function of a PID-control system is

$$G_{yr}(s) = \frac{N_P(s)N_C(s)}{sD_P(s) + N_P(s)N_C(s)} = \frac{1}{1 + \dfrac{sD_P(s)}{N_P(s)N_C(s)}}, \qquad (7.21)$$

where

$$\begin{cases} C(s) = K_P\left(1 + \dfrac{1}{T_i s} + T_d s\right) = \dfrac{N_C(s)}{s}, \ \ P(s) = \dfrac{N_P(s)}{D_P(s)}, \\ N_C(s) = K_i + K_P s + K_d s^2, \ \ K_i = K_P/T_i, \ \ K_d = K_P T_d. \end{cases} \qquad (7.22)$$

Let

$$\frac{sD_P(s)}{N_P(s)N_C(s)} = \frac{a_0 s + a_1 s^2 + a_2 s^3 + \cdots + a_n s^{n+1}}{\beta_0 + \beta_1 s + \beta_2 s^2 + \cdots + \beta_q s^q}$$
$$= \gamma_1 s + \gamma_2 s^2 + \gamma_3 s^3 + \cdots . \qquad (7.23)$$

Thus,

$$G_{yr}(s) = \frac{1}{1 + \Gamma(s)}, \ \ \Gamma(s) = \gamma_1 s + \gamma_2 s^2 + \gamma_3 s^3 + \cdots . \qquad (7.24)$$

We choose a desired input-output transfer function to be

$$G_m(s) = \frac{1}{1 + \Phi(s)} = \frac{1}{1 + \phi_1 \sigma s + \phi_2 \sigma^2 s^2 + \phi_3 \sigma^3 s^3 + \cdots}, \qquad (7.25)$$

where σ is a time-scaling parameter. We find the parameters of a PID
controller by choosing the poles of $G_{yr}(s)$ to be those of $G_m(s)$.

The parameters

$$\{\phi_1, \ \phi_2, \ \phi_3, \ \phi_4, \ \phi_5, \ \cdots\} = \left\{1, \ \frac{1}{2}, \ \frac{3}{20}, \ \frac{3}{100}, \ \frac{3}{1\,000}, \ \cdots\right\} \qquad (7.26)$$

provide us with a step response that minimizes an integral time absolute error

$$\text{ITAE} = \int_0^\infty t|e(t)|\mathrm{d}t \tag{7.27}$$

and yields an overshoot being about 10%. And the parameters

$$\{\phi_1, \ \phi_2, \ \phi_3, \ \phi_4, \ \cdots\} = \left\{1, \ \frac{3}{8}, \ \frac{1}{16}, \ \frac{1}{256}, \ \cdots\right\} \tag{7.28}$$

provide us with a fast step response that has a short settling time and does not have an overshoot.

We can find the PID parameters in (7.22) as follows.

Since we choose $\Gamma(s) = \Phi(s)$, $sD_P(s)/[N_P(s)N_C(s)] = \Phi(s)$. Thus,

$$N_C(s) = \frac{sH(s)}{\Phi(s)}, \quad H(s) = \frac{D_P(s)}{N_P(s)}, \tag{7.29}$$

Let

$$\begin{cases} D_P(s) = a_0 + a_1 s + a_2 s^2 + a_3 s^3 + \cdots + a_n s^n, \\ N_P(s) = b_0 + b_1 s + b_2 s^2 + b_3 s^3 + \cdots + b_m s^m, \end{cases}$$

we have

$$\begin{cases} H(s) = h_0 + h_1 s + h_2 s^2 + h_3 s^3 + \cdots + h_n s^n + \cdots, \\ h_0 = \dfrac{a_0}{b_0}, \ h_1 = \dfrac{a_1 - b_1 h_0}{b_0}, \ h_2 = \dfrac{a_2 - b_1 h_1 - b_2 h_0}{b_0}, \\ h_3 = \dfrac{a_3 - b_1 h_2 - b_2 h_1 - b_3 h_0}{b_0}, \\ h_i = \dfrac{a_i - b_1 h_{i-1} - b_2 h_{i-2} - \cdots - b_i h_0}{b_0}. \end{cases} \tag{7.30}$$

Substituting (7.30) into (7.29) and performing division for $N_C(s)$ yield

$$\begin{cases} K_i = \dfrac{h_0}{\phi_1 \sigma}, \ K_P = \dfrac{h_1 \phi_1 - \phi_2 h_0 \sigma}{\phi_1^2 \sigma}, \\ K_d = \dfrac{h_2 \phi_1^2 - \phi_2 \phi_1 h_1 \sigma + (\phi_2^2 - \phi_3 \phi_1) h_0 \sigma^2}{\phi_1^3 \sigma} \end{cases} \tag{7.31}$$

and

$$\begin{cases} \text{PI: } (\phi_2^2 - \phi_3\phi_1)h_0\sigma^2 - \phi_2\phi_1 h_1\sigma + h_2\phi_1^2 = 0, \\ \text{PID: } (\phi_2^3\phi_1 - 2\phi_3\phi_2 + \phi_4\phi_1)h_0\sigma^3 \\ \qquad + (\phi_3 - \phi_2^2)\phi_1 h_1\sigma^2 + \phi_2\phi_1 h_2\sigma - \phi_1^2 h_3 = 0. \end{cases} \qquad (7.32)$$

Thus, first, choose a desired $\Phi(s)$. Next, calculate h_i $(i = 1, 2, 3)$ in (7.30), choose the type of the controller (PI or PID), and solve the corresponding equation in (7.32). Finally, select the smallest positive σ (resulting in the fastest response) from the solutions and use it to calculate the controller parameters in (7.31).

Generally speaking, we can use the above methods to choose the parameters of a PID controller as initial values. Then, we carry out fine adjustment of those parameters using the pole-placement method and the dominant-pole method based on the time-domain specifications, T_r, M_p, and T_s. More specifically, we first specify the region for pole placement based on the time domain specifications. Then, we calculate the poles of the closed-loop control system based on the initial value of the PID parameters. Finally, we adjust the places of those poles with the consideration of dominant poles.

7.4   Two-Degree-of-Freedom Control System

The requirements for system design are

i)   the guarantee of the stability of a system

ii) satisfactory reference-tracking performance

iii) suppression of disturbance and uncertainties.

It is difficult to satisfy Requirements ii) and iii) simultaneously using a single controller in a feedback control system (Figure 7.3). Observing the configuration of the control system, we find that the big problem in the conventional feedback control system is that the system compresses the information. In fact, there are two available signals: $r(t)$ and $y(t)$, but the system compresses them to $e(t)$ and uses it to calculate the control input $u(t)$. This causes difficulty in system design.

There are two different kinds of signals in a control system: one is the exogenous signals, such as a reference input, a disturbance, and a noise; the other is measured signals, such as a displacement and a speed. If we can handle these two kinds of signals separately, we can alleviate the limitations in the system design and achieve a higher level of performance. For example, processing $r(t)$ and $y(t)$ separately

$$U(s) = \begin{bmatrix} C_1(s) & -C_2(s) \end{bmatrix} \begin{bmatrix} R(s) \\ Y(s) \end{bmatrix} = C_1(s)R(s) - C_2(s)Y(s) \quad (7.33)$$

yields a new configuration of a feedback control system (**Figure 7.6**). We call the configuration in Figure 7.3 a *one-degree-of-freedom control system* (1 自由度制御系) and that in Figure 7.6 a *two-degree-of-freedom control system* (2 自由度制御系)[†]. A simple calculation gives the transfer functions from the disturbance and the reference input to the output

$$G_{yd}(s) = \frac{P(s)}{1 + P(s)C_2(s)}, \quad G_{yr}(s) = \frac{P(s)C_1(s)}{1 + P(s)C_2(s)}. \quad (7.34)$$

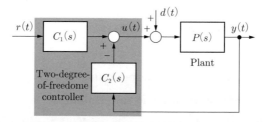

Figure 7.6   Two-degree-of-freedom control system

[†]   Since signals in a control system only have two kinds, the largest degrees of freedom for a control system is two.

We can use $C_2(s)$ to suppress the disturbance and use $C_1(s)$ to track the reference input. This enables the preferential adjustment of feedback and tracking performance.

7.5   Various PID-Control Systems

How to implement a designed PID controller is an important issue. We explain some well-known implementation strategies that have been developed to deal with problems in PID control.

(1) **Inexact differential**   An exact differentiator is not available in control practice. It also amplifies noise in a high-frequency band. Thus, we usually use an inexact differentiator

$$T_d s \doteqdot \frac{T_d s}{1 + \gamma T_d s}, \tag{7.35}$$

instead of the exact one and choose γ from the range $[0.1, 0.125]$.

(2) **PI-D control**   The control input produced by a derivative control term in a PID controller is

$$u_D(t) = K_P T_d \frac{\mathrm{d}e(t)}{\mathrm{d}t} = \begin{cases} K_P T_d \dfrac{\mathrm{d}r(t)}{\mathrm{d}t}, & t = 0, \\[2mm] -K_P T_d \dfrac{\mathrm{d}y(t)}{\mathrm{d}t}, & t > 0 \end{cases} \tag{7.36}$$

for a step reference input, $r(t)$, and $e(t) = r(t) - y(t)$. $\mathrm{d}r(t)/\mathrm{d}t|_{t=0} = \delta(t) = \infty$ but $\mathrm{d}r(t)/\mathrm{d}t = 0$ for $t > 0$. So, if we replace the differential of $e(t)$ into that of $y(t)$, it does not change the control effect but reduces vibrations. This gives the configuration in **Figure 7.7**. Since the differential action is carried out before the calculation of $e(t)$, thus the PI actions, this is called PI-D control.

(3) **Two-degree-of-freedom PID controller**   A feedforward-type two-degree-of-freedom PID control system is shown in **Figure 7.8**. A comparison between Figure 7.8 and Figure 7.6 gives

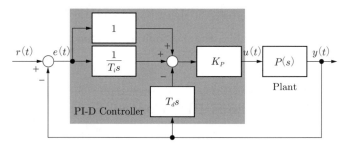

Figure 7.7   PI-D control system

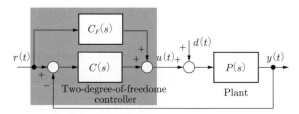

Figure 7.8   Two-degree-of-freedom PID control system

$$C_1(s) = C(s) + C_F(s), \quad C_2(s) = C(s). \tag{7.37}$$

$C_F(s)$ is used to improve tracking performance. Since a derivative control term quickens the response, a possible choice is

$$C_F(s) = -K_P(\alpha + \beta T_d s), \tag{7.38}$$

where $0 < \alpha, \beta < 1$. Some rules of selecting optimal α and β are given in Reference 12).

7.6   A Design Example

Consider the problem of designing a PI-speed-control system for the arm robot (3.27) in Chapter 3.

The plant is

$$P(s) = \frac{\beta}{s + \alpha}, \tag{7.39}$$

and the controller is

$$C(s) = K_P \left(1 + \frac{1}{T_i s} \right). \tag{7.40}$$

The input-output transfer function of the speed-control system is

$$G_{yr}(s) = \frac{\beta K_P \left(s + \dfrac{1}{T_i} \right)}{s^2 + (\alpha + \beta K_P)s + \dfrac{\beta K_P}{T_i}}. \tag{7.41}$$

Desired closed-loop poles p_1 and p_2 gives the characteristic equation of the closed-loop system

$$\phi(s) = (s - p_1)(s - p_2) = s^2 + \phi_1 s + \phi_0. \tag{7.42}$$

Comparing the denominator of $G_{yr}(s)$ with $\phi(s)$ yields

$$K_P = \frac{\phi_1 - \alpha}{\beta}, \quad T_i = \frac{\phi_1 - \alpha}{\phi_0}. \tag{7.43}$$

The following sample MATLAB commands carry out the analysis and design of the PI control system:

```
Program 7.1    PI controller for arm robot
1  s=tf('s');
2  % Plant
3  alpha=10; beta=50; P=beta/(s+alpha)
4  % Time-domain specifications
5  Tr=0.15; %[sec]
6  zeta1=0.7;
7  Ts=0.5; %[sec]
8
9  % Area for pole assignment
10 Omega1=1.8/Tr;
11 t1=pi/2:0.01:3*pi/2;
12 x1=Omega1*cos(t1); y1=Omega1*sin(t1);
13
14 K1=sqrt(1-zeta1^2)/zeta1;
```

```
15 t2=-12:0.1:0; yK1=K1*t2; yK2=-K1*t2;

16

17 sigma=4.6/Ts; ts=-12:0.1:12; ys=-sigma*ones(length(ts));

18

19 figure(1); hold on;

20 plot(x1,y1); plot(t2,yK1,t2,yK2); plot(ys,ts); sgrid;

21 title('Aera for pole placement')

22

23 % C(s)=Kp(1+1/(Tis))

24 % Pole assignment

25 p_FB=[-20+10*j -20-10*j];

26 plot(real(p_FB(1)), imag(p_FB(1)),'rx',real(p_FB(2)), imag(
     p_FB(2)),'rx');

27

28 phai=conv([1 -p_FB(1)],[1 -p_FB(2)]);

29 Kp=(phai(2)-alpha)/beta; Ti=Kp*beta/phai(3);

30 C=Kp*(1+1/(Ti*s))

31

32 L=C*P; figure(2); margin(L); [gm,PM,wcg,wcp]=margin(L)

33

34 G=L/(1+L); Gu=C/(1+L); Gr=minreal(G); figure(3); pzmap(Gr)

35

36 t=0:0.001:1;

37 figure(4); hold on;

38 step(Gr,t); step(Gu,t); grid;

39 title('Step response (Gyr & Gur)')

40 stepinfo(Gr)
```

The characteristics of a step response are displayed on the figure by right-clicking anywhere in the figure and selecting "Characteristics" from the pop-up menu (**Figure 7.9**).

The selected closed-loop poles for the design of the PI controller are shown in **Figure 7.10**(c). The Bode plots of the open-loop transfer func-

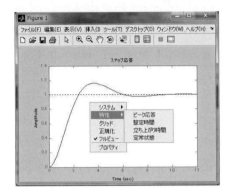

Figure 7.9 Display of characteristics of step response

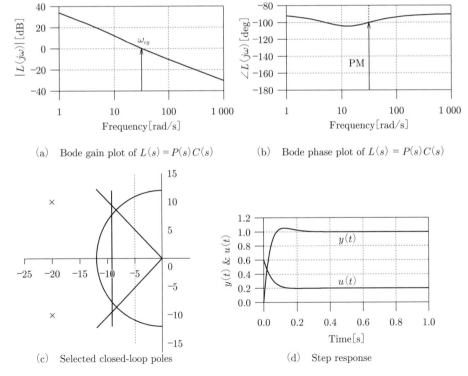

(a) Bode gain plot of $L(s) = P(s)C(s)$

(b) Bode phase plot of $L(s) = P(s)C(s)$

(c) Selected closed-loop poles

(d) Step response

Figure 7.10 Design results and verification

tion [Figures 7.10(a) and (b)] show that the gain and phase margins of the PI-control system are infinity and 79.9°, respectively. The step response and the corresponding control input are shown in Figure 7.10(d).

──────────── **Problems** ────────────

⟨ **Basic Level** ⟩

【1】 Verify that the relationships between the rise time T_r, the settling time T_s, and the time constant T for a first-order system $G(s) = K/(Ts+1)$ are $T_r = 2.2T$ and $T_s = 3T$ (for $\delta = 5\%$) [see (7.2)].

【2】 Let $P(s) = 6/[s(0.2s+1)(0.5s+1)]$ and $C(s) = K$ in a unity-feedback system.

    (1) Find gm and PM of each of the following systems:

        A. $K = 0.1$ B. $K = 0.2$ C. $K = 0.4$ D. $K = 0.6$.

    (2) As explained in Section 5.5, it is desirable that the PM of a system is larger or equal to 30°. Check if the statement is true for A∼D. And use Figure 6.9 to find the corresponding approximate ζ of each closed-loop system.

    (3) Compare the step responses of the closed-loop systems.

【3】 Let $P(s) = 1/[s(s+7)]$ and $C(s) = K$ in a unity-feedback system.

    A. Choose K that yields a step response with an overshoot of 15%.

    B. Evaluate the steady-state error for a unit ramp reference input for the designed K.

【4】 Let $G(s) = K/[(s+1)(s+3)(s+12)]$ in a unity-feedback system. Choose the gain K that yields a damping ratio of 0.8 for dominant poles. Compare the specifications of the uncompensated and compensated system.

【5】 Let $P(s) = 50/[s(s+10)]$ in a unity-feedback system. Use the pole-placement method to design the following controllers for control specifications: $T_r \leq 0.15\,\mathrm{s}$, $M_P \leq 10\%$, and $T_s \leq 0.5\,\mathrm{s}$ (beware of[†] choosing dominant and other poles). Carry out simulations for a step reference input and compare them.

    A. $C(s) = K_P(1+T_d s)$

────────────────

[†] beware of ∼ ： ∼に用心する。

B. $C(s) = K_P \left(1 + \dfrac{1}{T_i s} + T_d s\right).$

[6] Let $P(s) = 6/s(0.2s+1)(0.5s+1)$ and a PD controller be $C(s) = (0.4s+1)/(0.08s+1)$ in a unity-feedback system.

    A. Find how the controller $C(s)$ improves the gain and phase margins.

    B. Carry out simulations for the control system and the one without using the controller [that is, $C(s) = 1$]. Explain the effect of using the PD controller from the viewpoints of[†] the rise times and the overshoots.

[7] Let $P(s) = 50/[s(s+10)]$ and $C(s) = 10(1+T_d s)$ in a unity-feedback system.

    A. Find T_d such that the damping ratio of the closed-loop system is $\zeta = 0.707$.

    B. Carry out simulations for the control system and for the one that uses $C(s) = 1$. Explain the effect of using the inexact-differential controller $T_d s \doteqdot T_d s/(1 + 0.1 T_d s)$ from the viewpoints of the rise time and the overshoot.

[8] Consider a unity-feedback system in which the plant is $P(s) = K/[s(s+3)(s+6)]$ $(K > 0)$.

    A. Design a controller $C(s)$ that stabilizes the closed-loop system and drives the error of a ramp response to zero for an allowable K. Explicitly show the relationships between K and the parameters of $C(s)$.

    B. Choose K and $C(s)$ that satisfy the stability condition and carry out simulations to verify the designed results.

[9] Consider a two-degree-of-freedom control system (**Figure 7.11**) in which $P(s) = 1/[(0.1s+1)(0.5s+1)]$, $C(s) = K_P$, and $C_F(s) = K_F$.

    A. Find the range of K_P that guarantees the stability of the closed-loop system.

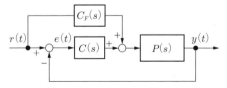

Figure 7.11 Problem 9

---

† from the viewpoint of $\sim$: 〜の観点から。

B.   Find the steady-state error for a unit step reference input when $C_F(s)$ is not used.

C.   Choose K_F that yields zero steady-state error for a step reference input.

Advanced Level

[1] Let the plant in a networked unity-feedback system be $P(s) = 2e^{-0.2s}/(s+2)$. Consider the following PI and PID design problems.

A.   Choose $n = 1$ and $m = 1$ and use Padé approximation for the time-delay term (Table 4.2). Find the approximated transfer function of $P(s)$, $\bar{P}(s)$.

B.   Compare the step responses for $P(s)$ and $\bar{P}(s)$.

C.   Use the partial-model-matching method to find a PI controller for $\{\phi_1, \phi_2, \phi_3\} = \{1, 1/2, 3/20\}$.

D.   Calculate the poles of the closed-loop control system for the PI controller and $\bar{P}(s)$.

E.   Use the unit step response to evaluate the designed PI controller.

[2] Consider the P-D control system in **Figure 7.12**. The derivative action in the inner-loop avoids a large overshoot. Design the inner-loop rate feedback that yields a damping ratio of 0.7 for the dominant poles of the inner-loop and a damping ratio of 0.5 for the dominant poles of the whole closed-loop system.

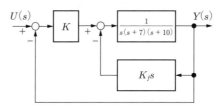

Figure 7.12   Advanced problem 2

8

Control-System Design in State Space

While a frequency-domain method (based on Laplace transform) is simple and easy to use, it can hardly provide us with the exact behavior of a system in the time domain during design. In this chapter, we explain a time-domain method. It focuses on the internal state of a system and tries to carry out precise control of a system exactly in the time domain. It has the following features:

i) Since it uses *simultaneous first-order differential equations* (連立一次微分方程式) to clearly describe the internal structure of a system, it eases the analysis of system characteristics, for example, it is simple to check whether or not we can control a specified variable in a system.

ii) We can use mathematical tools to easily check the stability of a system.

iii) The system description allows us to use a unified way to design not only a single-input single-output (SISO) system but also a multi-input multi-output (MIMO) system with coupled state components.

8.1 System Description

Let the transfer function of a system be

$$G(s) = \frac{b_{n-1}s^{n-1} + b_{n-2}s^{n-2} + \cdots + b_1 s + b_0}{s^n + a_{n-1}s^{n-1} + \cdots + a_1 s + a_0} \tag{8.1}$$

and the corresponding differential equation be

$$\frac{\mathrm{d}^n y(t)}{\mathrm{d}t^n} + a_{n-1}\frac{\mathrm{d}^{n-1}y(t)}{\mathrm{d}t^{n-1}} + \cdots + a_1 \frac{\mathrm{d}y(t)}{\mathrm{d}t} + a_0 y(t)$$
$$= b_{n-1}\frac{\mathrm{d}^{n-1}u(t)}{\mathrm{d}t^{n-1}} + \cdots + b_1 \frac{\mathrm{d}u(t)}{\mathrm{d}t} + b_0 u(t), \tag{8.2}$$

where $y(t)$ is the output and $u(t)$ is the input of the system. Rewriting (8.2) as

$$y^{(n)} = \underbrace{\left(-a_{n-1}y^{(n-1)} + b_{n-1}u^{(n-1)}\right) + \cdots + (-a_1\dot{y} + b_1\dot{u}) + \underbrace{\overbrace{(-a_0y + b_0u)}^{\dot{x}_1},}_{x_2^{(2)}}}_{x_n^{(n)}}$$

choosing the state variables, $x_1(t), x_2(t), \cdots, x_n(t)$, as shown in the above equation, and writing the equation as simultaneous first-order differential equations yiled

$$\begin{cases} \dot{x}_1(t) = -a_0 x_n(t) + b_0 u(t), \\ \dot{x}_2(t) = x_1(t) - a_1 x_n(t) + b_1 u(t), \\ \quad \vdots \\ \dot{x}_{n-1}(t) = x_{n-2}(t) - a_{n-2}x_n(t) + b_{n-2}u(t), \\ \dot{x}_n(t) = x_{n-1}(t) - a_{n-1}x_n(t) + b_{n-1}u(t), \\ y(t) = x_n(t), \end{cases} \tag{8.3}$$

where $\dot{\xi}(t) = \mathrm{d}\xi(t)/\mathrm{d}t$ and $\xi^{(i)} = \mathrm{d}^i\xi(t)/\mathrm{d}t^i$ $(i = 2, \cdots, n)$. Letting $x(t) = \begin{bmatrix} x_1(t), & x_2(t), & \cdots, & x_n(t) \end{bmatrix}^{\mathrm{T}}$ and representing (8.3) as a vector equation **(Figure 8.1)**

$$\begin{cases} \dot{x}(t) = Ax(t) + Bu(t), \\ y(t) = Cx(t) + Du(t), \end{cases} \tag{8.4}$$

we have

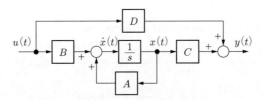

Figure 8.1 Block diagram of system (8.4)

$$\begin{cases} A = \begin{bmatrix} 0 & \cdots & 0 & -a_0 \\ 1 & & & -a_1 \\ & \ddots & & \vdots \\ & & 1 & -a_{n-1} \end{bmatrix}, \ B = \begin{bmatrix} b_0 \\ b_1 \\ \vdots \\ b_{n-1} \end{bmatrix}, \\ C = \begin{bmatrix} 0 & \cdots & 0 & 1 \end{bmatrix}, \ D = 0. \end{cases} \quad (8.5)$$

It gives a state-space realization of the system (8.2) (**Figure 8.2**), which is called the *observable canonical form* (可観測正準形).

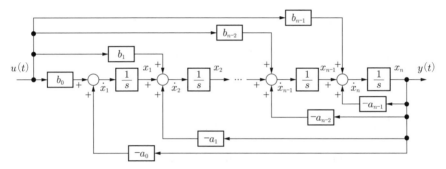

Figure 8.2 State-space realization of (8.2) in observable canonical form

On the other hand, if we define state variables to be

$$\begin{cases} \dot{x}_1(t) = x_2(t), \\ \dot{x}_2(t) = x_3(t), \\ \quad \vdots \\ \dot{x}_{n-1}(t) = x_n(t), \\ \dot{x}_n(t) = -a_0 x_1(t) - a_1 x_2(t) - \cdots - a_{n-1} x_n(t) + u(t), \\ y(t) = b_0 x_1(t) + b_1 x_2(t) + \cdots + b_{n-1} x_n(t), \end{cases} \quad (8.6)$$

we obtain another state-space realization of the system (8.2) (**Figure 8.3**), which is called the *controllable canonical form* (可制御正準形). If we write (8.6) as (8.4), then

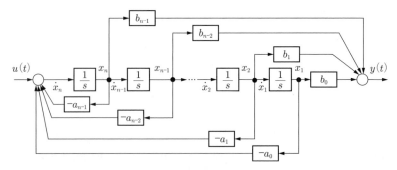

Figure 8.3   State-space realization of (8.2) in controllable canonical form

$$\begin{cases} A = \begin{bmatrix} 0 & 1 & & \\ \vdots & & \ddots & \\ 0 & & & 1 \\ -a_0 & -a_1 & \cdots & -a_{n-1} \end{bmatrix}, \ B = \begin{bmatrix} 0 \\ \vdots \\ 0 \\ 1 \end{bmatrix}, \\[20pt] C = \begin{bmatrix} b_0 & b_1 & \cdots & b_{n-1} \end{bmatrix}, \ D = 0. \end{cases} \qquad (8.7)$$

Solving the simultaneous first-order differential equations (8.4), we obtain the solution of the state

$$x(t) = e^{At}x(0) + \int_0^t e^{A(t-\tau)}Bu(\tau)\mathrm{d}\tau, \qquad (8.8)$$

where $x(0) = x(t)|_{t=0}$.

Applying the Laplace transform to (8.4) for zero initial condition $[x(0) = 0]$ yields the relationship between the state-space representation (A, B, C, D) and the transfer function of the sytem:

$$Y(s) = G(s)U(s), \ G(s) = D + C(sI - A)^{-1}B. \qquad (8.9)$$

As shown in (8.3) and (8.6), the state-space description of a system is not unique. Performing a *similarity transformation* (相似変換)

$$x(t) = T_S z(t) \qquad (8.10)$$

on the state (T_S is non-singular) yields a new description

$$\begin{cases} \dot{z}(t) = \bar{A}z(t) + \bar{B}u(t), \\ y(t) = \bar{C}z(t) + \bar{D}u(t), \end{cases} \tag{8.11}$$

$$\bar{A} = T_S^{-1}AT_S, \ \bar{B} = T_S^{-1}B, \ \bar{C} = CT_S, \ \bar{D} = D.$$

Note that

$$G(s) = \bar{D} + \bar{C}(sI - \bar{A})^{-1}\bar{B} = D + C(sI - A)^{-1}B,$$

the basic characteristics of the systems (8.4) and (8.11) are the same.

Since[†]

$$G(s) = D + C(sI - A)^{-1}B = D + \frac{C\mathrm{adj}(sI - A)B}{|sI - A|}, \tag{8.12}$$

the characteristic equation (5.19) is given by

$$|sI - A| = 0. \tag{8.13}$$

That is, the roots of (8.13) are exactly the poles of the system $G(s)$. Thus, the eigenvalues of A is exactly and the poles of $G(s)$.

8.2   Controllability and Observability

Controllability and observability are two fundamental concepts associated with a dynamical system.

The controllability of a system shows the ability to use the control signal, $u(t)$, to steer the behavior of all components of the state, $x(t)$. It is defined as follows.

8.2.1   Controllability

If we can find an input, $u(t)$, that transfers the state of the system (8.4)

[†]   adj(A) is the adjugate matrix of A.

from any initial state $x(t_0)$ to any desired final state $x(t_f)$ over a finite time interval $t_f - t_0 > 0$, then the system is said to be controllable; otherwise, the system is uncontrollable.

The controllability of the system (8.4) can easily be checked using a controllability matrix. Let $x(t) \in \mathbb{R}^n$. The controllability matrix is

$$M_C = \begin{bmatrix} B & AB & A^2 B & \cdots & A^{n-1} B \end{bmatrix}. \tag{8.14}$$

If the rank of M_C is n, the system (8.4) is controllable. Moreover, the implementation of a system in the controllable canonical form (8.7) ensures the controllability of the system.

8.2.2   Observability

The observability of a system shows the ability to observe state components from the output, $y(t)$. It is defined as follows.

If we can determine the initial state $x(t_0)$ from the time history of the input, $u(t)$, and the output, $y(t)$, over a finite time interval $t_f - t_0 > 0$, then the system is said to be observable; otherwise, the system is unobservable.

We can also construct an observability matrix

$$M_O = \begin{bmatrix} C \\ CA \\ CA^2 \\ \vdots \\ CA^{n-1} \end{bmatrix}. \tag{8.15}$$

If the rank of M_O is n, the system (8.4) is observable. Moreover, the implementation of a system in the obserable canonical form (8.5) ensures the observability of the system.

The example in **Figure 8.4** is used to illustrate the controllability and observability of a system. Since the system has two poles, its state has two

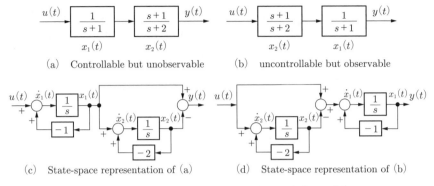

(a) Controllable but unobservable   (b) uncontrollable but observable

(c) State-space representation of (a)   (d) State-space representation of (b)

Figure 8.4 Examples of controllability and observability

components. Let $x_1(t)$ correspond to the pole $p_1 = -1$; and $x_2(t)$, $p_2 = -2$. Note that there is a pole-zero cancellation for $p_1 = -1$. Pole-zero cancellation results in uncontrollable or unobservable state components. First, we check the system in Figure 8.4 (a). The state $x_1(t)$ is directly controlled by $u(t)$ and it is then forwarded to $x_2(t)$. So, the system is controllable. However, since $x_1(t)$ is canceled by the zero located at the output, $x_1(t)$ cannot be observed from the output. Thus, the system is unobservable. Observation of the system in Figure 8.4 (b) in the same manner shows that the system is uncontrollable but observable. The following MATLAB commands verify the above statements:

Generally speaking, a system can be decomposed into four subsystems:

```
                Program 8.1    Verification of contr. and obs.

 1   Aa=[-1 0; 1 -2]; Ba=[1; 0]; Ca=[1 -1];
 2   Mca=ctrb(Aa,Ba); uncoa=length(Aa)-rank(Mca)
 3   Moa=obsv(Aa,Ca); unoba=length(Aa)-rank(Moa)
 4
 5   Ab=[-1 -1; 0 -2]; Bb=[1; 1]; Cb=[1 0];
 6   Mcb=ctrb(Ab,Bb); uncob=length(Ab)-rank(Mcb)
 7   Mob=obsv(Ab,Cb); unobb=length(Ab)-rank(Mob)
```

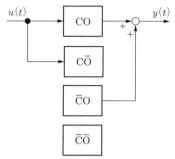

controllable and observable (CO), uncontrollable but observable ($\bar{\text{C}}$O), controllable but unobservable (C$\bar{\text{O}}$), and uncontrollable and unobservable ($\bar{\text{C}}\bar{\text{O}}$) ones (**Figure 8.5**). The transfer function only describes the controllable and observable subsystem.

Figure 8.5 System decomposition

As a result, we usually consider the problem of designing a control system for a plant that is controllable and observable, or in other words, we usually consider the problem of designing a controller for a CO subsystem. However, since the stability of a system is essential, the uncontrollable part is required to be stable for control. If a system meets this requirement, we say the system is stabilizable. In the same way, if the unobservable part of a system is stable, we say the system is detectable. Thus, we usually require that a system is stabilizable and detectable for control-system design.

8.3   Design of State Feedback

Consider the linear plant (8.4). The assumption that it is controllable guarantees that we can freely design a feedback gain to assign closed-loop poles to a state-feedback control system at any places in the s plane. We also assume that $D = 0$ for simplicity[†]. If all components of the state are available, we design a state-feedback gain

$$u(t) = -Fx(t) \tag{8.16}$$

[†] This system with $D = 0$ is called a strictly proper system. Most physical systems belong to this category.

to asymptotically stabilize the plant at the *origin* (原点). It is called the *regulator problem* (レギュレータ問題).

The closed-loop system becomes $\dot{x}(t) = (A - BF)x(t)$. We explain two widely used design methods for (8.16) below.

8.3.1   Pole-Placement Method

Let the feedback gain in (8.16) be

$$F = \begin{bmatrix} f_0, & f_1, & \cdots & f_{n-1} \end{bmatrix}. \tag{8.17}$$

If the plant is given in the controllable canonical form (8.7), then a simple calculation shows that

$$A - BF = \begin{bmatrix} 0 & 1 & & \\ \vdots & & \ddots & \\ 0 & & & 1 \\ -a_0 + f_0 & -a_1 + f_1 & \cdots & -a_{n-1} + f_{n-1} \end{bmatrix}, \tag{8.18}$$

and the characteristic equation of the close-loop system, $\phi(s) = |sI - (A - BF)| = 0$, is

$$s^n + (a_{n-1} - f_{n-1})s^{n-1} + \cdots + (a_1 - f_1)s + (a_0 - f_0) = 0. \tag{8.19}$$

On the other hand, if we choose the poles of the closed-loop system to be λ_i $(i = 1, 2, \cdots, n)$,

$$\phi(s) = \prod_{i=1}^{n}(s - \lambda_i) = s^n + \phi_{n-1}s^{n-1} + \cdots + \phi_1 s + \phi_0. \tag{8.20}$$

A comparison between (8.19) and (8.20) yields the state-feedback gain

$$f_i = a_i - \phi_i, \ i = 0, 1, 2, \cdots, n - 1. \tag{8.21}$$

If the system is not described in the controllable canonical form, then performing a state transformation (8.10) using

$$T_S = M_C \begin{bmatrix} a_1 & a_2 & \cdots & a_{n-1} & 1 \\ a_2 & a_3 & \cdots & 1 & 0 \\ \vdots & \vdots & \ddots & \ddots & \vdots \\ a_{n-1} & 1 & 0 & & \vdots \\ 1 & 0 & \cdots & \cdots & 0 \end{bmatrix} \tag{8.22}$$

yields the controllable canonical form. Note that M_C in (8.22) is the controllability matrix (8.14).

8.3.2 Linear-Quadratic Regulator

The most widely used method of designing a feedback gain is the *optimal control* (最適制御). Minimizing a quadratic performance index

$$J = \int_0^\infty \left\{ x^T(t)Qx(t) + u^T(t)Ru(t) \right\} dt \tag{8.23}$$

gives the optimal state-feedback gain

$$F = R^{-1}B^T P, \tag{8.24}$$

where P is the positive-definite symmetric solution of an algebraic Riccati equation

$$A^T P + PA - PBR^{-1}B^T P + Q = 0. \tag{8.25}$$

Q and R in (8.23) are weighting matrices. They are usually chosen to be positive-definite matrices ($Q > 0$ and $R > 0$). However, if we choose Q to be semi-positive definite ($Q \geqq 0$), then we have to ensure that ($Q^{1/2}, A$) is observable[†]. We choose a large entry on the main diagonal of Q to suppress the corresponding component of the state and use the same strategy to suppress control input.

The solution (8.24) is called the *linear-quadratic regulator* (LQR) (線形二次レギュレータ, 最適レギュレータ).

[†] If there exists a unique matrix M such that $Q = M^T M$, then $Q^{1/2} = M$.

8.4   State Observer

The state feedback (8.16) is used to construct a feedback control system. However, the state may not be available in control engineering. To solve this problem, we construct a state observer to estimate the state. For this purpose, we assume that (C, A) is observable. This assumption guarantees that we can freely assign poles to an observer of the plant.

We assume that $D = 0$ for simplicity and use the control input, $u(t)$, the output, $y(t)$, and the plant parameters (A, B, C) to build the dynamics of a state observer (**Figure 8.6**)

$$\begin{cases} \dot{\hat{x}}(t) = A\hat{x}(t) + Bu(t) + L[y(t) - \hat{y}(t)], \\ \hat{y}(t) = C\hat{x}(t), \\ \hat{x}(0) = 0 \end{cases} \tag{8.26}$$

to estimate the state of the plant (8.4). It is called a *full-order state observer* (同一次元オブザーバ). Some components of a state are usually available in many actual control systems. Thus, we may only need to build a state observer to estimate those unavailable state components. Such an observer is called a *reduced-order state observer* (低次元オブザーバ). Refer to other textbooks for this.

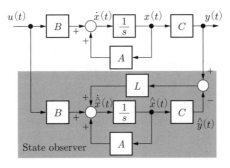

Figure 8.6 State observer

L in (8.26) is called the observer gain. Let $\Delta x(t) = x(t) - \hat{x}(t)$. A simple calculation gives

$$\Delta \dot{x}(t) = (A - LC)\Delta x(t), \ \Delta x(0) = x(0). \tag{8.27}$$

(8.27) shows that we can design L to adjust the convergence speed of the observer.

For a system (A, B, C, D), the system (A_d, B_d, C_d, D_d), in which

$$A_d = A^{\mathrm{T}}, \ B_d = C^{\mathrm{T}}, \ C_d = B^{\mathrm{T}}, \ D_d = D^{\mathrm{T}}, \tag{8.28}$$

is called the *dual system* (相対システム) of the system (A, B, C, D).

Note that the observability and controllability of a system are mathematical duals, that is, the controllability of a system (A, B, C, D) is exactly the observability of its dual system (A_d, B_d, C_d, D_d). The duality allows us to design the observer gain L by designing a state-feedback gain F_d for the dual system (A_d, B_d, C_d, D_d) and letting $L = F_d^{\mathrm{T}}$.

8.5   Observer-Based Control System

Combining the state produced by the state observer (8.26) and the state-feedback control law (8.16), we construct a feedback control system (**Figure 8.7**) that has the following state-space description

$$\begin{cases} \dfrac{\mathrm{d}}{\mathrm{dt}} \begin{bmatrix} x(t) \\ \Delta x(t) \end{bmatrix} = \begin{bmatrix} A & 0 \\ 0 & A - LC \end{bmatrix} \begin{bmatrix} x(t) \\ \Delta x(t) \end{bmatrix} + \begin{bmatrix} B \\ 0 \end{bmatrix} u(t), \\[2ex] y(t) = \begin{bmatrix} C & 0 \end{bmatrix} \begin{bmatrix} x(t) \\ \Delta x(t) \end{bmatrix}, \\[2ex] u(t) = \begin{bmatrix} -F & F \end{bmatrix} \begin{bmatrix} x(t) \\ \Delta x(t) \end{bmatrix}, \end{cases} \tag{8.29}$$

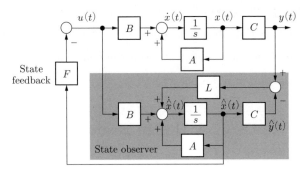

Figure 8.7   Regulator combined with state observer

Thus,

$$\frac{\mathrm{d}}{\mathrm{d}t}\begin{bmatrix} x(t) \\ \Delta x(t) \end{bmatrix} = \begin{bmatrix} A-BF & BF \\ 0 & A-LC \end{bmatrix}\begin{bmatrix} x(t) \\ \Delta x(t) \end{bmatrix}. \tag{8.30}$$

The characteristic equation of the closed-loop system is

$$\left| sI - \begin{bmatrix} A-BF & BF \\ 0 & A-LC \end{bmatrix} \right| = |sI-(A-BF)| \cdot |sI-(A-LC)| = 0.$$
$$\tag{8.31}$$

(8.31) shows that the poles of the closed-loop system are the combination of those of the regulator $(A-BF)$ and the observer $(A-LC)$. Thus, these two kinds of poles can be designed separately [the *separation theorem* (分離定理)].

There are no explicit rules to select the poles of an observer. We usually choose the poles of an observer in the left of those of the closed-loop state-feedback system in the s plane. This ensures that the state of the observer converges faster than that of the state-feedback system. In control engineering practice, many systems choose the real parts of the poles for an observer to be $2.5 \sim 10$ times the one of the dominant poles of the state-feedback system. Too left poles make an observer sensitive to[1] noise due to[2] its differential effect and also cause a big transient response. So, we need to consider the tradeoff between the convergent speed and noise reduction in the design of an observer.

8.6   A Design Example

Consider the arm robot in Examples 3.3 and 3.6 in Chapter 3 as an example.   Choosing state variables to be $x(t) = \begin{bmatrix} x_1(t) & x_2(t) \end{bmatrix}^T = \begin{bmatrix} \theta(t) & d\theta(t)/dt \end{bmatrix}^T$ and letting α and β be (3.27) allow us to write (3.21) as

$$\begin{cases} \dot{x}(t) = Ax(t) + Bu(t), \\ y(t) = Cx(t), \end{cases} \tag{8.32}$$

where

$$A = \begin{bmatrix} 0 & 1 \\ 0 & -\alpha \end{bmatrix}, \ B = \begin{bmatrix} 0 \\ \beta \end{bmatrix}, \ C = \begin{bmatrix} 1 & 0 \end{bmatrix}.$$

First, we assume that all the state components are available and consider the state-feedback control problem.

After checking the controllability of the plant, we choose the poles of the closed-loop system to be $-10 \pm j10$ (that is, the damping ratio of the closed-loop system is 0.707) to design a feedback gain. The MATLAB command `place` calculates the corresponding state-feedback gain as follows:

```
Program 8.2   Pole-placement
1  alpha=10; beta=50;
2  A=[0 1; 0 -alpha]; B=[0; beta]; C=[1 0];
3
4  Co=ctrb(A,B); unco=length(A)-rank(Co)
```

[1] be sensitive to ~ : ~に敏感である。
[2] due to ~ : ~のために、~の結果。

```
 5
 6  p=[-10+10*j -10-10*j];
 7  F=place(A,B,p);
 8  eig(A-B*F)
 9  sysF=ss(A-B*F,B,C,0); sysu=ss(A-B*F, B, -F, 1);
10
11  t=0:0.01:1;
12  figure(1)
13  subplot(2,1,1); impulse(sysF,t); grid;
14  subplot(2,1,2); impulse(sysu,t); grid;
```

Now, we consider the LQR design problem. We choose the weighting matrices Q and R to design a state-feedback control law (8.24). Note that the relative relationship between Q and R is key. Consider a scalar case as an example, the pairs ($Q = 10$, $R = 1$) and ($Q = 100$, $R = 10$) yield the same performance index and result in the same state-feedback gain. Another point is that, if we want to suppress a state component, we choose the corresponding weight to be a large value and vice versa.

We examine the following cases for the system design:

Case 1: $R = 1$ and $Q = \text{diag}\{1, 1\}$

Case 2: $R = 1$ and $Q = \text{diag}\{100, 1\}$

Case 3: $R = 10$ and $Q = \text{diag}\{100, 1\}$.

The purpose of the design is i) to compare Cases 1 and 2 to understand the effect of the weighting matrix Q and ii) to compare Cases 2 and 3 to understand the effect of the weight R. **Figure 8.8** shows the following facts:

i)   Since $y(t) = x_1(t)$, we can accelerate the output by increasing the (1, 1) entry in Q that corresponds to $x_1(t)$.

ii) Increasing R reduces the control input.

Sample MATLAB commands for the design are as follows:

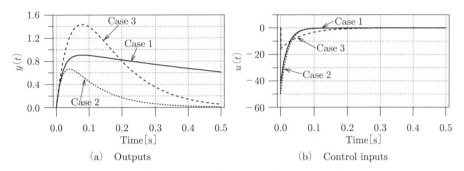

(a)   Outputs   (b)   Control inputs

Figure 8.8   Impulse responses for different
weightings, Q and R (Cases 1~3)

```
Program 8.3    LQR

 1  alpha=10; beta=50;
 2  A=[0 1; 0 -alpha]; B=[0; beta]; C=[1 0];
 3
 4  R=1; Q1=diag([1 10]);
 5  F1=lqr(A,B,Q1,R);
 6  eig(A-B*F1)
 7  sysF1=ss(A-B*F1,B,C,0); sysF1u=ss(A-B*F1, B, -F1, 1);
 8
 9  Q2=diag([100 1]);
10  F2=lqr(A,B,Q2,R);
11  eig(A-B*F2)
12  sysF2=ss(A-B*F2,B,C,0); sysF2u=ss(A-B*F2, B, -F2, 1);
13
14  R3=10;
15  F3=lqr(A,B,Q2,R3);
16  eig(A-B*F3)
17  sysF3=ss(A-B*F3,B,C,0); sysF3u=ss(A-B*F3, B, -F3, 1);
18
19  t1=0:0.01:0.5;
20  figure(1)
21  subplot(2,1,1); hold on;
22  impulse(sysF1,t); impulse(sysF2,t); impulse(sysF3,t); grid;
```

```
23  subplot(2,1,2); hold on;
24  impulse(sysF1u,t); impulse(sysF2u,t); impulse(sysF3u,t); grid;
```

Next, we assume that not all the state components are available. The position of the arm is measured by an encoder. We can use numerical differentiation to calculate the speed for ordinary positioning control. However, note that the speed resolution is one pulse per sampling time. For example, if we use an encoder of $1\,000\,\text{pulse}/360°$ and a sampling time of $0.001\,\text{s}$, then, the speed resolution is $(2\pi\,\text{rad}/1\,000\,\text{pulse})/0.001\,\text{s} = 2\pi\,\text{rad/s} = 60\,\text{rpm}^{\dagger}$. So, a positioning-control system with a low speed (less than $60\,\text{rpm}$ for this example) cannot use the numerical-differentiation method to obtain the speed). In this example, we design a full-order state observer.

After checking the observability of the plant, we choose the poles for the observer that meets the requirement given in the previous section, that is, their real parts are $2.5 \sim 10$ times the one of the dominant poles of the state-feedback system: $p_{1,2} = -30 \pm j10$. Then, we calculate an observer gain for the pair of the observer poles and check the impulse response of the observer (**Figure 8.9**).

The main MATLAB commands for the design of the observer are as follows:

Program 8.4   Observer design

```
1  Ob=obsv(A,C); unob=length(A)-rank(Ob)
2  p=[-30+10*j -30-10*j]; K=place(A',C',p); L=K'
3  eig(A-L*C)
4  sysOB=ss(A-L*C,B,C,0);
```

$\dagger$ The mainly used units of a rotational speed are (1) rad/s: radian per second, which is the SI unit; (2) rpm: revolutions per minute; and (3) and rps: revolutions per second. Note that rpm or rps are not SI units but they are accepted by the General Conference on Weights and Measures (CGPM) for use as being multiples or submultiples of SI units.

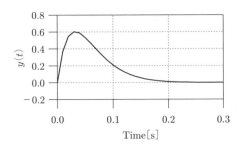

Figure 8.9 Impulse responses of observer

A comparison between the direct state-feedback control and the state-observer-based state-feedback control helps us understand the difference between these two systems and how should we design the state feedback and the observer (**Figure 8.10**). Some main MATLAB commands are given below.

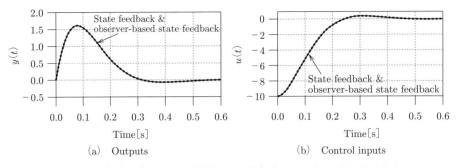

(a)   Outputs   (b)   Control inputs

Figure 8.10 Comparison of state feedback and
state-observer-based state feedback

```
         Program 8.5    State feedback w/ and w/o obs.

1   pf=[-10+10*i -10-10*i]; F=place(A,B,pf);

2   eig(A-B*F)

3   sysF=ss(A-B*F,B,C,0); sysFu=ss(A-B*F, B, -F, 1);

4

5   pob=[-30+10*j -30-10*j]; Kob=place(A',C',pob); L=Kob'

6   eig(A-L*C)
```

```
7  na=length(A); AFOB=[A-B*F B*F; zeros(na,na) A-L*C];
8  BFOB=[B; zeros(na,1)]; CFOB=[C zeros(1,na)];
9  sysFOB=ss(AFOB,BFOB,CFOB,0);
10 sysFOBu=ss(AFOB, BFOB, [-F F], 0);
11
12 t=0:0.01:0.6;
13 figure(1)
14 subplot(2,1,1); impulse(sysF,sysFOB,t); grid;
15 subplot(2,1,2); impulse(sysFu,sysFOBu,t); grid;
```

Problems

Basic Level

[1] Name two advantages of feedback control.

[2] For a system

$$
\begin{cases}
\dot{x}(t) = \begin{bmatrix} 2 & 1 & 1 \\ 1 & 7 & 1 \\ -3 & 4 & -5 \end{bmatrix} x(t) + \begin{bmatrix} 0 \\ 0 \\ 1 \end{bmatrix} r(t), \\
y(t) = \begin{bmatrix} 0 & 1 & 0 \end{bmatrix} x(t),
\end{cases}
\tag{8.33}
$$

find out how many poles are on the $j\omega$ axis, in the left-half, and in the right-half s plane.

[3] Determine whether or not the following system is controllable

$$
\dot{x}(t) = \begin{bmatrix} -1 & 1 & 2 \\ 0 & -1 & 5 \\ 0 & 3 & -4 \end{bmatrix} x(t) - \begin{bmatrix} 2 \\ 1 \\ 1 \end{bmatrix} u(t).
$$

[4] Consider a system represented in the state space

$$
A = \begin{bmatrix} 0 & 1 & -1 \\ -2 & -3 & 0 \\ a & 1 & 1 \end{bmatrix}, \ B = \begin{bmatrix} 1 \\ 0 \\ 0 \end{bmatrix}, \ C = \begin{bmatrix} 0 & 0 & 1 \end{bmatrix}, \ D = 0.
$$

A. Find the range of a ensuring the controllability of the system.

B.   Find the range of a ensuring the observability of the system.

C.   Find the range of a ensuring the system stability.

D.   Find the transfer function of the system when the system is controllable and observable, uncontrollable but observable, and controllable but unobservable.

[5]   The dynamics of an artificial satellite in an orbit above the equator are

$$\begin{cases} \dot{x}(t) = Ax(t) + Bu(t), \\ y(t) = Cx(t), \\ x(t) = [r(t),\ \dot{r}(t),\ \theta(t),\ \dot{\theta}(t)]^{\mathrm{T}}, \\ u(t) = [u_r(t),\ u_\theta(t)]^{\mathrm{T}}, \\ y(t) = [r(t),\ \theta(t)]^{\mathrm{T}}, \\ A = \begin{bmatrix} 0 & 1 & 0 & 0 \\ 3\omega^2 & 0 & 0 & 2\omega \\ 0 & 0 & 0 & 1 \\ 0 & -2\omega & 0 & 0 \end{bmatrix}, \quad B = \begin{bmatrix} 0 & 0 \\ b_r & 0 \\ 0 & 0 \\ 0 & b_\theta \end{bmatrix}, \quad C = \begin{bmatrix} 1 & 0 & 0 & 0 \\ 0 & 0 & 1 & 0 \end{bmatrix}, \end{cases}$$

where ω is the angular speed at which the satellite moves around the earth, $r(t)$ is the distance between the satellite and the earth, $\theta(t)$ is the angle between the satellite and the equatorial plane, $u_r(t)$ is the driving force on the diameter direction, and $u_\theta(t)$ is the driving force on the tangent direction.   Examine the controllability of the system when $u_r(t)$ is used ($b_r = 1$ and $b_\theta = 0$), $u_\theta(t)$ is used ($b_r = 0$ and $b_\theta = 1$), and both are used ($b_r = 1$ and $b_\theta = 1$).

[6]   Consider a linear system

$$\begin{cases} \dot{x}(t) = \begin{bmatrix} 0 & 1 \\ -2 & -3 \end{bmatrix} x(t) + \begin{bmatrix} 0 \\ 1 \end{bmatrix} u(t), \\ y(t) = \begin{bmatrix} 1 & 2 \end{bmatrix} x(t). \end{cases}$$

If we use a state feedback $u(t) = -Fx(t)$ ($F = [f_1\ f_2]$). For what combination of f_1 and f_2 the closed-loop system is unstable?

[7]   Let $m = 1\,\mathrm{kg}$, $l = 1\,\mathrm{m}$, $J = 0.25\,\mathrm{kg \cdot m^2}$, and $b = 0.01\,\mathrm{Nm \cdot s/rad}$, and $\theta_0 = 0$. Consider the problem of designing a state-feedback control law for an arm robot on the vertical plane (3.46).

A.   Use the pole-placement method to place the poles of the closed-loop system at $-10 \pm j5$.

B. Use LQR with $Q = \text{diag}\{100,\ 1\}$ and $R = 1$.

C. Draw the Bode plots of the closed-loop system for the two design methods.

D. Calculate the impulse responses for the two systems.

[8] Find the controllable and observable canonical forms of the arm robot $P(s) = 50/[s(s + 10)]$.

[9] Consider a plant $P(s) = (s + 2)/[(s + 5)(s + 9)]$.

A. Find the controllable canonical form of the plant.

B. Design an observer for the plant with a transient response close to a second-order system with $\zeta = 0.6$ and $\omega_n = 120\,\text{rad/s}$.

[10] Choose poles to be $\{-10,\ -20\}$ to construct an observer for a system $\dot{x}(t) = \begin{bmatrix} 0 & 1 \\ 0 & 0 \end{bmatrix} x(t) + \begin{bmatrix} 0 \\ 1 \end{bmatrix} u(t)$ and $y(t) = \begin{bmatrix} 1 & 0 \end{bmatrix} x(t)$.

[11] Design an observer for the plant $G(s) = 1/[(s + 7)(s + 8)(s + 9)]$, whose state-space representation is given by

$$A = \begin{bmatrix} -7 & 1 & 0 \\ 0 & -8 & 1 \\ 0 & 0 & -9 \end{bmatrix},\ B = \begin{bmatrix} 0 \\ 0 \\ 1 \end{bmatrix},\ C = \begin{bmatrix} 1 & 0 & 0 \end{bmatrix}.$$

Find the gain of an observer that has a step response with an overshoot of 10% and a settling time of 0.1 s.

[12] A large antenna system is $P(s) = 10/[s(s + 5)(s + 10)]$.

A. Find the controllable canonical form of $P(s)$.

B. Use the LQR method to design an observer for $P(s)$.

C. Use the LQR method to design a state-feedback gain for $P(s)$.

D. Find the impulse time response of the designed observer-based state-feedback control system.

Advanced Level

[1] Consider a linear system

$$\begin{cases} \dot{x}(t) = \begin{bmatrix} 0 & 0 & 0 \\ 1 & a & 0 \\ 0 & 0 & b \end{bmatrix} x(t) + \begin{bmatrix} 1 \\ 0 \\ c \end{bmatrix} u(t), \\ y(t) = \begin{bmatrix} 0 & 1 & d \end{bmatrix} x(t), \end{cases}$$

where a, b, c, and d are real numbers.

    A. Find the condition that the system is controllable but not observable, observable but not controllable, and both controllable and observable.

    B. Find the input-output transfer functions when the system is controllable but not observable, observable but not controllable, and both controllable and observable.

    C. Find if the system is asymptotically stable.

[2] An astronaut operates a satellite-recovery arm robot in a spacecraft to collect satellites. The transfer function of a robot is $P(s) = e^{-0.01s}/[s(s+10)]$.

    A. Use a Padé approximation with $m = n = 2$ in Table 4.2 to find an approximate transfer function of the robot, $\bar{P}(s)$.

    B. Find the controllable canonical form of $\bar{P}(s)$.

    C. Use the LQR method to design an observer for $\bar{P}(s)$.

    D. Use the LQR method to design a state-feedback control law for $\bar{P}(s)$.

    E. Find the impulse response of the system.

9 | Controller Implementation

Implementation of a controller is important in control engineering practice. While we usually design a controller in continuous time, we mainly use a microprocessor to implement it in discrete time. This chapter explains some basic issues related to the implementation of controllers.

9.1 Procedure of System Implementation

The design and implementation of a control system have four steps: plant modeling, system analysis, controller design, and controller implementation. There are four procedures to carry out the design and implementation: Procedures A~D (**Figure 9.1**). Procedure A performs all steps in the continuous-time domain. However, since a microprocessor is widely used in control engineering practice nowadays, Procedure A is not very common nowadays. On the other hand, Procedure D performs all steps in the discrete-time domain and Procedure C performs the discretization of a plant at the first step and other steps in the discrete-time domain. Since the relationship between poles and a time response is not intuitively understandable in the discrete-time domain, Procedure B, which designs a continuous controller and discretizes it to produce a discrete-time controller, is the most widely used one.

Another point in the system design is that we need to derive a low-order

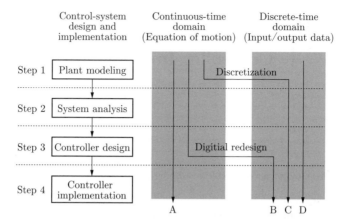

Figure 9.1 Procedures of control system design and implementation

controller that eases the implementation.

9.2   Model Reduction

The order of a system becomes high if we try to build a precise model. Control-system design based on a high-precision model usually results in a high-order controller. Generally speaking, simulations based on high-order models are precise, and real-time control using high-order controllers yields required control performance. On the other hand, the implementation of a high-order dynamic system not only is computationally expensive but also may decrease system reliability.

Model reduction contains plant-model reduction and controller reduction. The procedure for reducing the order of a model is shown in **Figure 9.2**. We do not distinguish between a plant

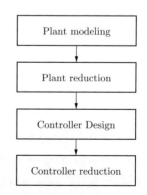

Figure 9.2 Procedure for model reduction

and[†] a controller hereafter because the basic considerations for model reduction are the same.

The most important thing for model reduction is that we need to use a reduced-order controller to verify the performance of the closed-loop control system for the original plant.

9.2.1　Treatment of Small Time Constants

Neglecting or combining small time constants is a widely used method of reducing the order of a model. While it is a little rough, it is practical.

First, we compare the step responses of the transfer functions in each of the two cases:

$$\text{Case 1}: G_{11}(s) = \frac{1}{s+1},\ G_{12} = \frac{1}{(0.1s+1)(s+1)}; \tag{9.1}$$

$$\text{Case 2}: G_{21}(s) = \frac{1}{s^2+s+1},\ G_{22} = \frac{1}{(0.1s+1)(s^2+s+1)}. \tag{9.2}$$

Program 9.1　Comp. of step resp.

```
1  s=tf('s');
2
3  G11=1/(s+1); G12=1/(0.1*s+1)*G11;
4  G21=1/(s^2+s+1); G22=1/(0.1*s+1)*G21;
5
6  t=0:0.01:10;
7  figure(1); hold on;
8  step(G11,t); step(G12,t); grid;
9  title('Step response (G11 & G12)')
10
11 figure(2); hold on;
12 step(G21,t); step(G22,t); grid;
13 title('Step response (G21 & G22)')
```

---

† distinguish between A and B：A と B を区別する。

The step responses of $G_{11}(s)$ and $G_{12}(s)$, and $G_{21}(s)$ and $G_{22}(s)$ are very close (**Figure 9.3**). Or in other words, we can consider $G_{11}(s)$ $[G_{21}(s)]$ as a reduced model of $G_{12}(s)$ $[G_{22}(s)]$. The pole plots of these transfer functions show that, except for the same poles that $G_{11}(s)$ and $G_{12}(s)$ $[G_{21}(s)$ and $G_{22}(s)]$ have, $G_{12}(s)$ $[G_{22}(s)]$ has another pole far from the imaginary axis in the left-half s plane that corresponds to a small time constant. The dynamics of such a time constant is very fast. Thus, the behavior of a system is mainly determined by dominant poles (the poles close to the imaginary axis) and the effect of the small time constant can be neglected. Now, the problem is that we need to define a criterion to check whether or not time constants are small enough that can be neglected.

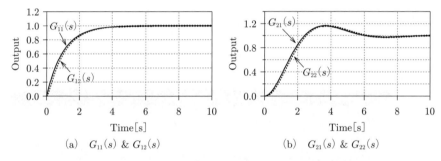

(a) $G_{11}(s)$ & $G_{12}(s)$ (b) $G_{21}(s)$ & $G_{22}(s)$

Figure 9.3   Step responses of $G_{11}(s)$, $G_{12}(s)$, $G_{21}(s)$, and $G_{22}(s)$

Consider a transfer function

$$G(s) = \frac{K(1 + T_1 s)}{s^2(1 + T_2 s)(1 + T_3 s)(1 + T_4 s) \cdots (1 + T_n s)} \qquad (9.3)$$

that has the cutoff angular frequency ω_c. If time constants T_i $(i = 3, 4, \cdots, n)$ satisfy

$$\sum_{i=3}^{n} T_i \le \frac{0.1}{\omega_c}, \qquad (9.4)$$

then we say that the time constants T_i $(i = 3, 4, \cdots, n)$ are small. We can

(**1**)　neglect them and reduce the original system (9.3) to

$$G(s) = \frac{K(1 + T_1 s)}{s^2(1 + T_2 s)},\tag{9.5}$$

(**2**)　combine them as $T_\Sigma = \sum\limits_{i=3}^{n} T_i$ and reduce the original system (9.3) to

$$G(s) = \frac{K(1 + T_1 s)}{s^2(1 + T_2 s)(1 + T_\Sigma s)}.\tag{9.6}$$

The Bode magnitude plots of (9.3), (9.5), and (9.6) are shown in **Figure 9.4**.

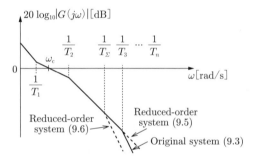

Figure 9.4　Bode magnitude plots of system before and after reduction

9.2.2　Modal-Decomposition Method

Let the roots of a characteristic equation $D(s) = 0$ be λ_i ($i = 1, 2, \cdots, n$) (see Chapter 5). The exponentials $e^{\lambda_i t}$ ($i = 1, 2, \cdots, n$) are called the *characteristic modes* (固有モード) (also called modes or natural modes). The modal-decomposition method reduces the order of a system by neglecting high-frequency modes.

Let a system be

$$G(s) = \frac{b_m s^m + b_{m-1} s^{m-1} + \cdots + b_1 s + b_0}{s^n + a_{n-1} s^{n-1} + \cdots + a_1 s + a_0}, \quad m \leqq n.\tag{9.7}$$

Carrying out partial-fraction decomposition for (9.7) yields

$$G(s) = K + \left(\frac{\beta_1}{s + \alpha_1} + \cdots + \frac{\beta_r}{s + \alpha_r} \right) + \left[\frac{\beta_{r+1}}{s + \alpha_{r+1}} + \cdots + \frac{\beta_n}{s + \alpha_n} \right].$$

(9.8)

If we can divide the poles of the system into two clusters (**Figure 9.5**) and the cluster contains the poles $-\alpha_i$ $(i = r + 1, \cdots, n)$ are in a region that is much farther from the imaginary axis than the one contains the poles $-\alpha_j$ $(j = 1, \cdots, r)$, then the modes in this group can be neglected and a reduced-order model is

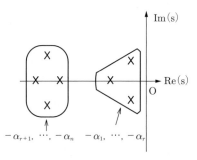

Figure 9.5   Grouping of poles

$$G(s) = K + \left(\frac{\beta_1}{s + \alpha_1} + \cdots + \frac{\beta_r}{s + \alpha_r} \right) + \left[\frac{\beta_{r+1}}{\alpha_{r+1}} + \cdots + \frac{\beta_n}{\alpha_n} \right]. \quad (9.9)$$

Note that we need to ensure that the DC gains of the system before and after model reduction are the same. Thus, we have to reserve the items in the brackets in (9.9).

[Example 9.1]   Use the modal-decomposition method to carry out model reduction for

$$G(s) = \frac{1}{(s^2 + s + 1)(0.01s + 1)}.$$

The partial-fraction decomposition for the system gives

$$G(s) = \frac{-0.0101s + 0.9999}{s^2 + s + 1} + \frac{0.0101}{s + 100}. \quad (9.10)$$

Since the mode corresponding to the pole, $s + 100 = 0$, is much farther from the imaginary axis in the left-half plane than the modes corresponding to the poles $s^2 + s + 1 = 0$, we neglect this mode. As a result, the reduced-order model is

$$G_r(s) = \frac{0.0001s^2 - 0.01s + 1}{s^2 + s + 1}. \tag{9.11}$$

As shown in this example, a reduced-order model derived by this method may produce new zeros that may result in uncontrollable or unobservable modes. So, we need to check the characteristics of the system after model reduction (**Figure 9.6**).

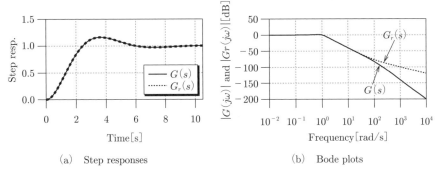

(a)   Step responses   (b)   Bode plots

Figure 9.6   Comparison of models for modal-decomposition method

The MATLAB commands for the example are as follows.

```
     Program 9.2   Modal decomposition
 1  s=tf('s');

 2

 3  G=1/((s^2+s+1)*(0.01*s+1))
 4  a=conv([1 1 1],[0.01 1]); b=[0 0 0 1];

 5

 6  [beta,alpha,k]=residue(b,a)

 7

 8  Gr=beta(2)/(s-alpha(2))+beta(3)/(s-alpha(3))-beta(1)/
        alpha(1)
 9  RG=beta(1)/(s-alpha(1))

10

11  sys_G=ss(Gr)
12  Co=ctrb(sys_G); unco=length(sys_G.A)-rank(Co)
```

```
13  Ob=obsv(sys_G); unob=length(sys_G.A)-rank(Ob)
14
15  figure(1);
16  step(G,Gr); grid;
17  title('Step responses of G and Gr')
18
19  figure(2);
20  bode(G,Gr); grid;
21  title('Bode plots of G and Gr')
```

9.3 z Transform

In a computer-controlled system, a continuous-time signal $f(t)$ $(t \geqq 0)$ is sampled to produce a discrete-time sequence $f[i]$ $(i = 0, 1, 2, \cdots)$ (**Figure 9.7**). While the Laplace transform

$$F(s) = \int_0^\infty f(t)e^{-st}\mathrm{d}t \tag{9.12}$$

deals with a continuous-time signal, the z transform

$$F(z) = \sum_{i=0}^\infty f[i]z^i \tag{9.13}$$

handles a discrete-time signal, where i indicates the time points $t = ih$ $(i = 0, 1, 2, \cdots)$ and h is a *sampling period* (サンプリング周期).

These two operators are connected by $z = e^{hs}$. z is called a *forward-shift operator* (進み演算子)[†] that shifts a signal one step ahead in time:

[†] q is usually used to denote the forward-shift operator. z is, in fact, a complex variable. The difference is that the z-transform considers the initial value of a signal but the q operator does not. However, since the initial value of a signal is usually taken to be zero in the pulse-transfer-function-based analysis, we ignore the difference for simplicity in this book.

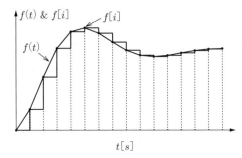

Figure 9.7 Continuous-time signal $f(t)$ and
its sampled version $f[i]$

$zx[i] = x[i + 1]$. Conversely, $z^{-1} = e^{-hs}$ is a *backward-shift operator* (遅れ
演算子) that shifts a signal one step back in time: $z^{-1}x[i] = x[i - 1]$.

9.4 System Description in Discrete-Time Domain

Assume that a sampling period h is used in a computer-controlled system
(**Figure 9.8**). An analog-to-digital (A/D) converter samples the output
of a plant at sampling instants $t = ih$ for $i = 0, 1, 2, \cdots$. It produces
a sequence $y[i]$ (a discrete-time signal) that represents the values of the
output $y(t)$. A *zero-order hold* (ゼロ次ホールド) is used as a mathematical
model of a digital-to-analog (D/A) converter that converts a discrete-time
signal $u[i]$ to a continuous-time one:

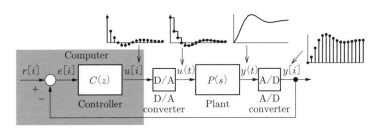

Figure 9.8 Computer-controlled system

$$u(t) = u[i], \ ih \le t < (i+1)h. \tag{9.14}$$

While a system is described by a transfer function in the continuous-time domain [for example, $P(s)$ in Figure 9.8], it is also described by a pulse-transfer function in the discrete-time domain [for example, $C(z)$ in Figure 9.8].

Assume that a continuous-time system is

$$\begin{cases} \dot{x}(t) = Ax(t) + Bu(t), \\ y(t) = Cx(t) + Du(t). \end{cases} \tag{9.15}$$

The solution of the state is

$$x(t) = e^{A(t-t_0)}x(t_0) + \int_{t_0}^{t} e^{A(t-\tau)}Bu(\tau)\mathrm{d}\tau. \tag{9.16}$$

Note that (9.14) holds between sampling instants. If we choose $t_0 = ih$ and $t = (i+1)h$, then we obtain a discrete-time state-space equation of the system

$$\begin{cases} x[i+1] = \Phi x[i] + \Gamma u[i], \\ y[i] = Cx[i] + Du[i], \end{cases} \tag{9.17}$$

where

$$\Phi = e^{Ah}, \ \Gamma = \int_{0}^{h} e^{A\tau}\mathrm{d}\tau \ B.$$

Thus, we have the pulse-transfer function

$$G(z) = D + C(zI - \Phi)^{-1}\Gamma \tag{9.18}$$

of the system, which is the counterpart of the transfer function

$$G(s) = D + C(sI - A)^{-1}B \tag{9.19}$$

in the continuous-time domain.

In addition to[†] the above precise description in the discrete-time domain, some other approximate descriptions are also widely used. Assume that the sampling period h is small enough. We use a *forward difference* (前進差分) to approximate a derivative (Euler's method):

$$\frac{dx}{dt} \doteq \frac{x[i+1] - x[i]}{h} = \frac{z-1}{h}x[i]$$

$$\Longleftrightarrow \quad z = e^{hs} \doteq 1 + hs \quad \left(s \doteq \frac{z-1}{h} \right) \tag{9.20}$$

or the *backward difference* (後退差分) method to approximate a derivative:

$$\frac{dx}{dt} \doteq \frac{x[i] - x[i-1]}{h} = \frac{1-z^{-1}}{h}x[i]$$

$$\Longleftrightarrow \quad z = \frac{1}{e^{-hs}} \doteq \frac{1}{1 - hs} \quad \left(s \doteq \frac{z-1}{hz} \right). \tag{9.21}$$

Another approximation is Tustin transform

$$z = e^{hs} = \frac{e^{hs/2}}{e^{-hs/2}} \doteq \frac{1 + hs/2}{1 - hs/2} \quad \left(s \doteq \frac{2}{h}\frac{z-1}{z+1} \right). \tag{9.22}$$

A continuous-time system is stable if the poles of the system are in the open left-half s plane. A discrete-time system is stable if the poles of the system are inside the unit disk in the z plane (**Figure 9.9**). Likewise, a continuous-time system is a minimum-phase one if both poles and zeros of the system are in the left-half s plane. A discrete-time system is a

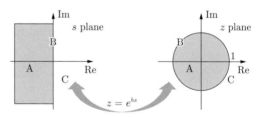

Figure 9.9 Mapping regions of s plane onto z plane

minimum-phase one if both poles and zeros of the system are inside the unit disk in the z plane. Tustin transform preserves the stability and minimum-phase property.

[†] in addition to ~：～のほかに。

Euler's method, the backward difference, and Tustin transform are special cases of the bilinear transform

$$s = \frac{\alpha z + \delta}{\gamma z + \beta} \tag{9.23}$$

that transforms a continuous-time system (A, B, C, D) to a discrete-time system

$$\begin{cases} A_d = (\beta A - \delta I)(\alpha I - \gamma A)^{-1}, \ B_d = (\alpha\beta - \gamma\delta)(\alpha I - \gamma A)^{-1}B, \\ C_d = C(\alpha I - \gamma A)^{-1}, \ D_d = D + \gamma C(\alpha I - \gamma A)^{-1}B. \end{cases}$$

$$\tag{9.24}$$

9.5   Selection of Sampling Period

The behavior of a computer-controlled system approaches that of a continuous-time control system as a sampling period goes to zero. While a small sampling period suppresses the loss of information by sampling, it imposes a heavy computational burden on a controller. A large sampling period reduces the computational expense at the cost of[†1] information loss. Thus, how to choose a sampling period is an important issue. A tradeoff between control performance, computational expense, and stability should be taken into careful consideration[†2].

Shannon's sampling theorem gives the condition for sampling and reconstruction of a continuous-time signal.

Shannon's sampling theorem   Consider a continuous-time signal $f(t)$ with its components in an angular frequency band $[-\omega_f, \omega_f]$ rad/s. It is possible to reconstruct $f(t)$ from its sampled version $f[i]$ if and only if the sampling angular frequency is higher than $2\omega_f$.

---

†1   at the cost (expense) of $\sim$ ： $\sim$を犠牲にして。
†2   take $\sim$ into consideration ： $\sim$を考慮に入れる，斟酌する。

This theorem provides us with a rationale for the selection of a sampling period. Based on the above theorem, we know that, if we choose the sampling period to be h [s], that is, the sampling angular frequency is $\omega_s = 2\pi/h$ [rad/s], we can only perfectly reconstruct a signal that has components with angular frequencies lower than $\omega_N = \omega_s/2$, where ω_N is called the *Nyquist angular frequency* (ナイキスト角周波数). If a signal has components with frequencies higher than ω_N, then we need to use a low-pass filter, which is called an antialiasing filter, to remove the high-frequency components.

The sampling period given by Shannon's sampling theorem should be considered as an upper bound. It is usually chosen to be much smaller than that given by the theorem. A proper choice of a sampling period in a computer-controlled system depends on the properties of a signal, the purpose of the system, and many other factors. Some rules of thumb are explained below.

If we focus on the rise time of a continuous-time plant, T_r, it is reasonable to choose the sampling period that satisfies

$$T_r/h = 4 \sim 10. \tag{9.25}$$

If the natural angular frequency of the plant is ω_n, then

$$h = (0.2 \sim 0.6)\frac{1}{\omega_n} \tag{9.26}$$

might be another possible choice that has 10 to 40 samples per vibration period. Since a time response to a disturbance may be different from that to a reference input for a high-order plant, the sampling period may need to be adjusted for a closed-loop system that contains such a plant.

The sampling period should be selected such that the control performance of a closed-loop control system is not influenced by sampling. Let the bandwidth of the closed-loop system be ω_b. The sampling period is usually chosen such that the sampling angular frequency is 6 to 10 times larger than ω_b[†1], that is,

$$h = \frac{2\pi}{(6 \sim 10)\omega_b} = (0.6 \sim 1)\frac{1}{\omega_b}. \tag{9.27}$$

If the desired natural angular frequency of the closed-loop control system is ω_n,

$$h = (0.1 \sim 0.6)\frac{1}{\omega_n} \tag{9.28}$$

is also another reasonable choice.

Note that the sampling period selected for a control system is large, compared to that used in signal processing. The reason is that the dynamics of many plants are of a low-pass character and the rise times of those plants are usually larger than those of their closed-loop control systems.

9.6   System Implementation

Many devices can be used in cooperation with Simulink Real-Time Toolbox[†2] to perform real-time control for a designed control system. Speedgoat[†3] and sBOX II[†4] are designed for this purpose. However, they are expensive and not suitable for personal use.

To lower the barrier for the users of this book, we build a control system using the combination of an Arduino[†5] (**Figure 9.10**) and MATLAB/

[†1]   Åström and Wittenmark recommended that the sampling angular frequency is 10 to 30 times larger than ω_b [15].

[†2]   We need to use Simulink to carry out system implementation. However, a detailed explanation is omitted due to space restrictions. Readers are recommended to refer to related books.

[†3]   https://www.mttis.co.jp/items/model_base_design/speedgoat/

[†4]   https://www.mttis.co.jp/items/model_base_design/s-box2/

[†5]   https://jp.mathworks.com/hardware-support/arduino-matlab.html

Simulink. An Arduino is a low-cost single-board microcontroller that contains a hardware board (Arduino AVR) and a software development environment (Arduino IDE). There are two ways to use it:

Figure 9.10 Arduino board

i) One is to generate codes from a Simulink model and then to directly execute them on the Arduino (it is called *run on target hardware*).

ii) The other is to use the ArduinoIO Package that considers an Arduino as an input/output device. This allows us to treat an Arduino and a plant connected to the Arduino as one box in a Simulink file.

Since ii) is easy to use, we recommend it for the readers of this book. To carry out experiments, we first need to install MATLAB/Simulink on a host PC and use a USB cable to connect the PC with an Arduino.

Note that, since we do not use Real-Time Toolbox in the system, the system cannot carry out exact real-time control but only pseudo-real-time control, that is, sampling periods may not be always precisely the same due to multi-tasks performed in the Windows system. Moreover, the sampling rate cannot be very high because serial communication is used between an Arduino and a host PC.

The steps of experimental preparation are

Step 1 : Download the Arduino IDE[†] and install it on a host PC.

Step 2 : Connect the Arduino to the host PC using a USB cable, open the control panel of the PC, left-click "Hardware and sound" and then

---

† https://www.arduino.cc/en/software

"Device manager". Find an unknown device and right-click the icon to update the driver. Finally, confirm the number of a serial port assigned to the Arduino.

Step 3 : Download and install ArduinoIO from Mathworks[†], execute the file `install_arduino.m`, run Arduino IDE, and execute the file `adiosrv.pde`.

A two-mass system (**Figure 9.11**) is a model of the steering system of a car, coupled oscillators, and many other mechatronic systems. In Figure 9.11(b), the variables and parameters with the subscript p (or d)

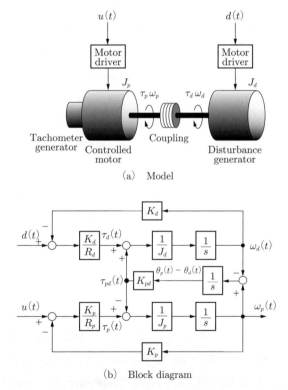

(a)   Model

(b)   Block diagram

Figure 9.11   Two-mass system

---

†   https://jp.mathworks.com/matlabcentral/fileexchange/32374

indicate that they are related to the controlled motor (or disturbance generator). $u(t)$ is the applied voltage [V], $d(t)$ is a disturbance voltage [V], $i_p(t)$ $[i_d(t)]$ is the armature current [A], $\tau_p(t)$ $[\tau_d(t)]$ is the torque produced [Nm], $\tau_{pd}(t)$ is the twisting torque [Nm], $\omega_p(t)$ $[\omega_d(t)]$ is the rotational speed [rad/s], $\theta_p(t)$ $[\theta_d(t)]$ is the rotation angle [rad], J_p (J_d) is the inertia [kgm^2], R_p (R_d) is the resistance of armature coil [Ω], K_p (K_d) is the back-electromotive-force constant [Vs/rad] or torque constant [Nm/A], and K_{pd} is the twisting elasticity coefficient of coupling [Nm/rad]. Choosing the state to be $x(t) = [\omega_p(t) \; \omega_d(t) \; \theta_p(t) - \theta_d(t)]^{\mathrm{T}}$ and the output to be $y(t) = \omega_p(t)$ yields the following state-space description:

$$
\begin{cases}
A = \begin{bmatrix} -\dfrac{K_p^2}{J_p R_p} & 0 & -\dfrac{K_{pd}}{J_p} \\[2mm] 0 & -\dfrac{K_d^2}{J_d R_d} & \dfrac{K_{pd}}{J_d} \\[2mm] 1 & -1 & 0 \end{bmatrix}, \quad B = \begin{bmatrix} \dfrac{K_p}{J_p R_p} \\[2mm] 0 \\[2mm] 0 \end{bmatrix}, \\[10mm]
B_d = \begin{bmatrix} 0 \\[1mm] \dfrac{K_d}{J_d R_d} \\[1mm] 0 \end{bmatrix}, \quad C = \begin{bmatrix} 1 & 0 & 0 \end{bmatrix}.
\end{cases}
$$

An experimental set [21)] verifies the speed control of a two-mass system. The parts of the system (**Figure 9.12**) are available at an electric store.

Arduino IO Setup and Real-Time Pacer must be placed in a Simulink model to ensure that an experiment is carried out properly. The Simulink model

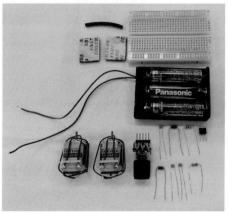

Figure 9.12 Experimental elements for motor positioning control

and experimental results of the PID positioning control of the two-mass system are shown in **Figure 9.13**.

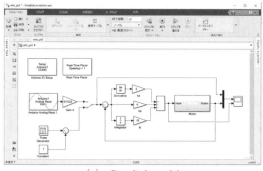

(a)   Simulink model

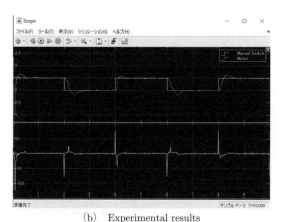

(b)   Experimental results
(upper part: input and output, lower part: control input)

Figure 9.13   Experimental results of PID speed control

See Hirata's book [21] for detailed instruction on how to use an Arduino board and MATLAB/Simulink to construct experimental systems.

——————— **Problems** ———————

Basic Level

【1】   Which of the following devices converts a continuous-time signal to a sequence of pulses?

   A. ZOH   B. D/A converter   C. A/D converter.

[2] Choose a device to decode a numerically coded signal to a continuous-time signal from the following list:

   A. A/D converter   B. D/A converter   C. ZOH.

[3] Choose a mathematical model for a digital-to-analog converter that reconstructs a continuous-time signal from a discrete-time sequence

   A. D/A converter   B. A/D converter   C. ZOH.

[4] We usually ignore the inductance of an armature circuit for a small-size DC motor because an electrical time constant $(T_e = L_a/R_a)$ is much smaller than a mechanical time constant $[T_m = JR_a/(K_T K_E)]$ for those motors. On the other hand, we have to consider the effect of an inductance for a medium- or large-size DC motor. A mathematical model of a DC motor (**Figure 9.14**) is given by

$$\left\{ \begin{array}{l} \text{Armature circuit: } u(t) = R_a i_a(t) + L_a \dfrac{\mathrm{d}i_a(t)}{\mathrm{d}t} + e_g(t), \\[2mm] \text{Fleming's left-hand rule: } \tau_M(t) = K_T i_a(t), \\[2mm] \text{Fleming's right-hand rule: } e_g(t) = K_E \omega(t), \\[2mm] \text{Motion: } J\dfrac{\mathrm{d}\omega(t)}{\mathrm{d}t} + b\omega(t) = \tau_M(t), \end{array} \right. \qquad (9.29)$$

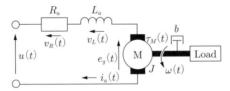

Figure 9.14 Permanent-magnet DC motor

where L_a [H] is the inductance of the armature circuit and J [kg $\cdot$ m^2] is the moment of inertial of the motor and load. Other parameters are the same as those given in Example 3.3 in Chapter 3. Consider the problem of speed control and solve the following problems:

   A. We let $b = 0$ for simplicity. Verify that the characteristic equation of the motor is $T_e T_m s^2 + T_m s + 1 = 0$.

   B. Let the poles of the motor be s_1 and s_2. If one pole, say, s_2, is more than 5 times farther from the imaginary axis than the other one, we can consider that, compared to the time constant corresponding to s_1, that corresponding to s_2 is small. So, we can ignore it and build a reduced-order system that only has one pole, s_1. Find the condition that we can ignore the electrical time constant T_e.

C. Consider a 350-W small-size DC motor: $R_a = 0.13\,\Omega$, $L_a = 0.0001\,\text{H}$, $K_T = K_E = 0.076$, $b = 0.00036\,\text{Nms/rad}$, and $J = 0.04\,\text{kgm}^2$. Check if we can ignore the electrical time constant and find a reduced-order system. And draw the Bode plots of the systems before and after reduction to verify the decision.

D. Consider a 37-kW medium-size DC motor: $R_a = 0.08\,\Omega$, $L_a = 0.004\,\text{H}$, $K_T = K_E = 1.18$, $b = 0.001\,\text{Nms/rad}$, and $J = 0.51\,\text{kgm}^2$. Check if we can ignore the electrical time constant and find a reduced-order system. And draw the Bode plots of the systems to verify the decision.

【5】 Use MATLAB command c2d to convert a continuous-time controller $C(s) = 1/(s^2 + 2s + 2)$ to a z transfer function for a sampling period of 0.005 s.

【6】 Use a ZOH (9.14) and Tustin transform (9.22) (MATLAB command: c2d and 'tustin') with the sampling period of 0.001 s to find the PD and PID controllers (use inexact differential for D actions): $C(s) = 5(1 + 0.08s)$ and $C(s) = 35[1 + 1/(0.14s) + 0.04s]$. Use Simulink to carry out simulations

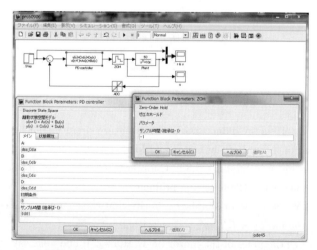

Figure 9.15　Implementation of PD controller

for the unit step reference input and compare the output responses (**Figure 9.15**).

Advanced Level

[1] Consider a motor control system in a machine tool (**Figure 9.16**). The transfer function of the motor is $P(s) = 5/(0.012s^2 + 0.097s + 1)$, that of the driver is $G_d(s) = 30e^{-0.0017s}$, and the gain of the speed sensor is $K_s = 0.015$. Design a PID controller $C(s) = K_P[1 + 1/(T_i s) + T_d s]$ for the speed-control system:

    A. Calculate the transfer function of the plant $G_P(s) = K_s P(s) G_d(s)$.

    B. Use Padé approximation with $n = 1$ and $m = 0$ for the time delay in $G_d(s)$. Draw the Bode plots and check whether or not (9.4) holds so as to find a reduced-order system. If it is, compare the Bode plots of the original plant and the reduced one.

    C. Find a $C(s)$ that yields $T_r = 0.1$ s, $M_P = 10\%$, and $T_s = 0.3$ s for the reduced-order systems given in B.

    D. Carry out simulations to compare the step responses using the designed controllers for the original and the reduced-order plants.

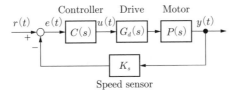

Figure 9.16 Motor control system

[2] Use Simulink to examine the implementations of a controller $C(s) = 1 + 1/s$ in a unity-feedback system in which $P(s) = 50/s(s + 10)$. Note that an AD converter is implemented with a `Quantizer` in the folder "Discontinuities"; and a DA converter, a `Zero-Order-Hold` in the folder "Discrete" in Simulink (for example, **Figure 9.17**).

    A. Use Simulink to construct a simulator for the original continuous-time system.

    B. Use Simulink to construct a simulator that uses Euler's method to

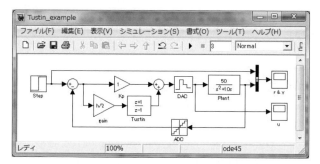

Figure 9.17   Implementation of controller using
Tustin transform

build a discrete-time controller.

C.   Use Simulink to construct a simulator that uses the backward difference method to build a discrete-time controller.

D.   Use Simulink to construct a simulator that uses Tustin transform to build a discrete-time controller.

E.   Set `quantization interval` in a `Quantizer` to be 0.01. Choose the sampling period to be $h = 0.001$, 0.005, 0.01, 0.05, 0.1, 0.5, and 1.0 s. Carry out simulations, compare the outputs and control inputs, and find a suitable range of sampling periods for the implementation of this system.

F.   Set the sampling period to be $h = 0.001$ s. Choose `quantization intervals` in `Quantizer`s for AD and DA converters separately to be 0.001, 0.005, 0.01, 0.05, 0.1, 0.5, and 1.0, that is, fix a `Quantizer` for an AD converter and choose a `Quantizer` for a DA converter successively from the above values and vice versa. Carry out simulations, compare the outputs and control inputs, and find a suitable quantization interval for AD and DA converters.

10

Comprehensive Exercises

The final chapter uses two numerical examples to show how we can use control theory to design control systems that provide us with satisfactory control performance. The arm-robot control illustrates transfer-function-based control-system design. And the control of the pendulum-and-cart system illustrates state-space-based control system design.

10.1 Arm-Robot Control

We design a PID controller

$$C(s) = K_P \left(1 + \frac{1}{T_i s} + T_d s \right) \tag{10.1}$$

to carry out the positioning control of the arm robot

$$P(s) = \frac{\beta}{s(s + \alpha)}, \ \alpha = 10, \ \beta = 50. \tag{10.2}$$

The transfer function from the reference input to the output is

$$G_{yr}(s) = \frac{P(s)C(s)}{1 + P(s)C(s)} = \frac{\beta K_P \left(T_d s^2 + s + \frac{1}{T_i} \right)}{s^3 + (\alpha + \beta K_P T_d)s^2 + \beta K_P s + \frac{\beta K_P}{T_i}}. \tag{10.3}$$

To start with[†], we find the ranges of the control parameters to guarantee the stability of the closed-loop system. Routh's stability criterion tells us

---

[†] to start with : まず第一に，最初に (= first of all)。

that all the coefficients of the characteristic polynomial should be positive, that is,

$$\alpha + \beta K_P T_d > 0, \ \beta K_P > 0, \ \frac{\beta K_P}{T_i} > 0. \tag{10.4}$$

Then, we construct a Routh table and obtain the following condition

$$\beta K_P(\alpha + \beta K_P T_d) > \frac{\beta K_P}{T_i}. \tag{10.5}$$

Combining (10.4) and (10.5) yields the range of the control parameters:

$$K_P > 0, \ T_i > 0, \ T_d > \frac{1 - \alpha T_i}{\beta T_i K_P}, \tag{10.6}$$

that is, T_d should be above the curved surface in **Figure 10.1**.

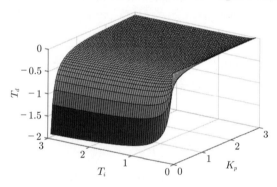

Figure 10.1   Ranges for PID parameters for (10.2)

If we choose the characteristic polynomial to be

$$\phi(s) = s^3 + \phi_2 s^2 + \phi_1 s + \phi_0, \tag{10.7}$$

then the parameters of the PID controller satisfy

$$\alpha + \beta K_P T_d = \phi_2, \ \beta K_P = \phi_1, \ \frac{\beta K_P}{T_i} = \phi_0. \tag{10.8}$$

As a result, we obtain

$$K_P = \frac{\phi_1}{\beta}, \ T_i = \frac{\phi_1}{\phi_0}, \ T_d = \frac{\phi_2 - \alpha}{\phi_1}. \tag{10.9}$$

We choose the closed-loop poles for the same control specifications given in Section 7.6 ($T_r = 0.15$ s, $\zeta = 0.7$, and $T_s = 0.5$ s) to be $\{-20\pm j10, \ -50\}$. They give the PID controller

$$K_P = 50, \ T_i = 0.1000, \ T_d = 0.0320. \tag{10.10}$$

These parameters satisfy Condition (10.6) shown in Figure 10.1.

We use an inexact differentiator to implement the differentiator. The MATLAB commands for the design and a step response are as follows.

Program 10.1   PID controller for arm robot

```
1  % Plant
2  s=tf('s');
3  alpha=10; beta=50; P=beta/(s*(s+alpha))
4
5  % Pole assignment and PID parameters
6  p_FB=[-20+10*j -20-10*j -50];
7  phai=conv(conv([1 -p_FB(1)],[1 -p_FB(2)]), [1 -p_FB(3)]);
8  Td=(phai(2)-alpha)/phai(3)
9  Kp=phai(3)/beta
10 Ti=phai(3)/phai(4)
11
12 C=Kp*(1+1/(Ti*s)+Td*s/(0.1*Td*s+1))
13 L=C*P,
14 G_PID=L/(1+L); G_PIDr=minreal(G_PID)
15 G_PIDu=C/(1+L); G_PIDur=minreal(G_PIDu)
16
17 t=0:0.001:1;
18 subplot(2,1,1);
19 step(G_PIDr);
20 title('Output of step response')
21 subplot(2,1,2);
22 step(G_PIDu,t);
23 title('Control input of step response')
24
25 z = zero(G_PIDr), p = pole(G_PIDr)
```

The step responses show that, while the output quickly tracks the reference input, it has a large overshoot and the control input has a large peak at the beginning of the step response (**Figure 10.2**). As explained in Chapter 7, the reason is that the PID controller adds two zeros $(-15.625 \pm j8.268)$ to the system.

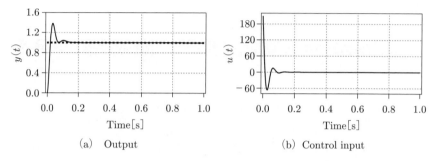

(a)   Output   (b)   Control input

Figure 10.2   Step response of PID

Adopting PI-D control (see Section 7.5 for details) is a strategy that reduces the overshoot because the control system carries out the differential action before PI actions. The input-output transfer function is

$$G_{yr}(s) = \frac{\beta K_P \left(s + \dfrac{1}{T_i} \right)}{s^3 + (\alpha + \beta K_P T_d)s^2 + \beta K_P s + \dfrac{\beta K_P}{T_i}}. \tag{10.11}$$

That is, compared to PID control, PI-D control has only one zero. We use the PI-D control scheme to suppress the overshoot of the output and the peak of the control input. Simulation results (**Figure 10.3**) show that the control input is successfully suppressed by this strategy, but there is still a visible overshoot.

Finally, we use a two-degree-of-freedom control system structure to further improve tracking performance. We select a reference model, which describes a desired input-output characteristic, to be

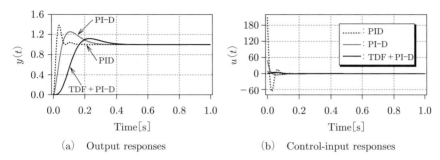

(a) Output responses   (b) Control-input responses

Figure 10.3 Step responses of PID, PI-D,
and two-degree-of-freedom PI-D control

$$G_M(s) = \frac{\omega_M^2}{s^2 + 2\zeta_M \omega_M s + \omega_M^2}. \tag{10.12}$$

Since the poles $-20 \pm j10$ were chosen for the desired closed-loop system,
we choose the characteristic polynomial of the model to be

$$s^2 + 2\zeta_M \omega_M s + \omega_M^2 = (s + 20 - j10)(s + 20 + j10) = s^2 + 40s + 500 \tag{10.13}$$

and use it to shape the reference input. The output of the model, $r_M(t)$, is
added to the control system (**Figure 10.4**). Since $r_M(t)$ is smoother than
the step single, the output of the system has a smaller overshoot than that
of the PI-D control system (Figure 10.3).

It is worth mentioning that the overshoot is reduced at the expense of[†]

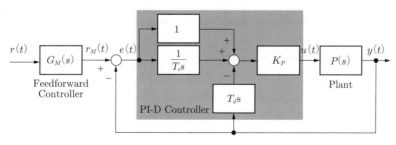

Figure 10.4 Configuration of two-degree-of-freedom PI-D control

---

[†] at the expense (cost) of ～ : ～を犠牲にして。

the rise time.

10.2   Pendulum-and-Cart Control

Consider the problem of designing a control system for the pendulum-and-cart system explained in Section 3.3 in Chapter 3. Assume that $\theta(t)$ is small enough such that the linear approximation model (3.58) is applicable. Let the state be $x(t) = \begin{bmatrix} z(t) & \theta(t) & \dot{z}(t) & \dot{\theta}(t) \end{bmatrix}^{\mathrm{T}}$ and the linear approximation model of (3.58) be

$$
\begin{cases}
\dfrac{\mathrm{d}x(t)}{\mathrm{d}t} = A_P x(t) + B_P f_c(t), \\
y(t) = C_P x(t), \\
A_P = \begin{bmatrix} 0 & 0 & 1 & 0 \\ 0 & 0 & 0 & 1 \\ 0 & -2 & -6 & 0.002 \\ 0 & 40 & 6 & -0.04 \end{bmatrix}, \; B_P = \begin{bmatrix} 0 \\ 0 \\ 16 \\ -15 \end{bmatrix}, \\
C_P = \begin{bmatrix} 1 & 0 & 0 & 0 \\ 0 & 1 & 0 & 0 \end{bmatrix}.
\end{cases}
\tag{10.14}
$$

First, we check the controllability and observability of the plant to make sure[†] that the plant is controllable and observable.

```
   Program 10.2   Controllability and observability
1  % Plant
2  ap=[0 0 1 0; 0 0 0 1; 0 -2 -6 0.002; 0 40 6 -0.04];
3  bp=[0; 0; 16; -15]; cp=[1 0 0 0; 0 1 0 0]; dp=[0; 0];
4
5  Co=ctrb(ap,bp); unco=length(ap)-rank(Co)
6  Ob=obsv(ap,cp); unob=length(ap)-rank(Ob)
```

---

† make sure ~：～を確かめる，確実に～する。

The controllability of the plant ensures that we can find a control law to perform a suitable control for the cart position and the pendulum angle. While large weights for $z(t)$ and $\theta(t)$ result in quick responses, they cause a large control input. Trial and error gives

$$Q = \text{diag}\{1,\ 5,\ 0.1,\ 0.1\},\ R = 1. \tag{10.15}$$

They yield an acceptable impulse response (**Figure 10.5**).

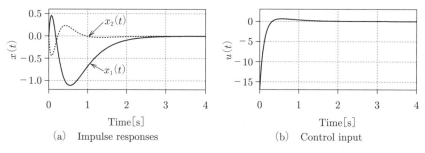

(a)   Impulse responses   (b)   Control input

Figure 10.5   Impulse response of LQR control for pendulum and cart

```
    Program 10.3   LQR for pendulum-and-cart
1  R=1; Q=diag([1 5 0.1 0.1]);
2  F=lqr(ap,bp,Q,R);
3
4  sys=ss(ap-bp*F,bp,cp,dp); sysu=ss(ap-bp*F,bp,-F,0);
5
6  figure(1)
7  impulse(sys); grid;
8  figure(2)
9  impulse(sysu); grid;
```

Next, we design a tracking controller to ensure that the cart position tracks a given step signal

$$r(t) = 1(t). \tag{10.16}$$

We choose the new output to be

$$y_1(t) = C_{P1}x(t), \quad C_{P1} = \begin{bmatrix} 1 & 0 & 0 & 0 \end{bmatrix}. \tag{10.17}$$

Simple verification shows that the plant is also controllable and observable. The internal model in the control system is

$$A_R = 0, \quad B_R = 1. \tag{10.18}$$

To design a servo system (**Figure 10.6**), we construct an augmented-state representation containing the plant and the internal model of the step signal

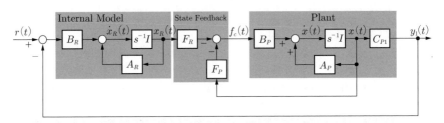

(a)   State feedback

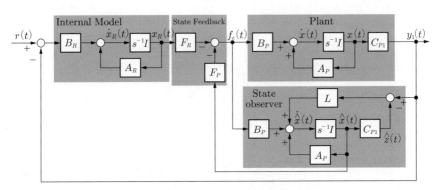

(b)   State-observer-based state feedback

Figure 10.6   Servo system for pendulum-and-cart system

$$
\begin{cases}
\dfrac{\mathrm{d}\bar{x}(t)}{\mathrm{d}t} = \bar{A}\bar{x}(t) + \bar{B}f_c(t) + \bar{B}_R r(t), \\[2mm]
y_1(t) = \bar{C}\bar{x}(t), \\[2mm]
f_c(t) = -\bar{F}\bar{x}(t), \\[2mm]
\bar{x}(t) = \begin{bmatrix} x_R(t) \\ x(t) \end{bmatrix}, \ \bar{A} = \begin{bmatrix} A_R & -B_R C_{P1} \\ 0 & A_P \end{bmatrix}, \\[4mm]
\bar{B} = \begin{bmatrix} 0 \\ B_P \end{bmatrix}, \ \bar{B}_R = \begin{bmatrix} B_R \\ 0 \end{bmatrix}, \\[4mm]
\bar{C} = \begin{bmatrix} 0 & C_{P1} \end{bmatrix}, \ \bar{F} = \begin{bmatrix} F_R & F_P \end{bmatrix}.
\end{cases}
\tag{10.19}
$$

The augmented system $(\bar{A}, \ \bar{B}, \ \bar{C})$ needs to be controllable and observable to guarantee that we can design a suitable control law for the system. Note that the steady state of $x(t)$ is not zero. The performance index (8.23) cannot be used directly because the integrated value will tend to infinity. Thus, we define

$$
\begin{aligned}
\delta x(t) &= x(t) - x(+\infty) \\
\delta x_R(t) &= x_R(t) - x_R(+\infty) \\
\delta f_c(t) &= f_c(t) - f_c(+\infty).
\end{aligned}
\tag{10.20}
$$

and minimize the following performance index

$$
J_F = \int_0^\infty \left\{ \begin{bmatrix} \delta x_R^{\mathrm{T}}(t) & \delta x^{\mathrm{T}}(t) \end{bmatrix} Q_F \begin{bmatrix} \delta x_R(t) \\ \delta x(t) \end{bmatrix} + R_F \delta f_c^2(t) \right\} \mathrm{d}t.
\tag{10.21}
$$

The selection of

$$
Q_F = \mathrm{diag}\{1, \ 10, 10, \ 1, \ 1\}, \ R_F = 1
$$

yields

$$
[F_R \mid F_P] = [1.0000 \mid -3.8791 \ \ -33.6043 \ \ -3.2280 \ \ -5.4735].
$$

The step response of the servo system with the state feedback is shown in **Figure 10.7**.

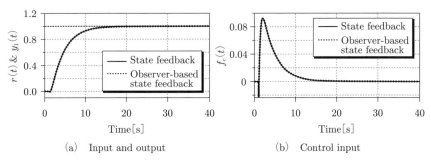

(a) Input and output   (b) Control input

Figure 10.7 Step response of servo system with with state feedback and with state-observer-based state feedback

However, if the state of the plant is not available, we need to design a state observer to carry out state feedback. Let the state observer be

$$
\begin{cases}
\dfrac{d\hat{x}(t)}{dt} = A_P\hat{x}(t) + B_P f_c(t) + LC_{P1}[x(t) - \hat{x}(t)], \\
\hat{y}_1(t) = C_{P1}\hat{x}(t).
\end{cases}
\tag{10.22}
$$

Since the poles of the servo system are -0.3181, -3.0113, $-4.8933 \pm j1.8160$, and -23.3787, we choose the poles of the observer to be $-10 \pm j2$ and $-20 \pm j5$, which are more than thirty times farther from the imaginary axis in the left-half s plane than the dominant pole of the closed-loop system, -0.3181. The resulting servo system is

$$
\begin{cases}
\dfrac{d\tilde{x}(t)}{dt} = \tilde{A}\tilde{x}(t) + \tilde{B}f_c(t) + \tilde{B}_R r(t), \\[2mm]
y_1(t) = \tilde{C}\tilde{x}(t), \\[2mm]
f_c(t) = -\tilde{F}\tilde{x}(t), \\[2mm]
\tilde{x}(t) = \begin{bmatrix} x_R(t) \\ x(t) \\ \hat{x}(t) \end{bmatrix}, \quad
\tilde{A} = \begin{bmatrix} A_R & -B_R C_{P1} & 0 \\ 0 & A_P & 0 \\ 0 & LC_{P1} & A_P - LC_{P1} \end{bmatrix}, \\[6mm]
\tilde{B} = \begin{bmatrix} 0 \\ B_P \\ B_P \end{bmatrix}, \quad
\tilde{B}_R = \begin{bmatrix} B_R \\ 0 \\ 0 \end{bmatrix}, \\[6mm]
\tilde{C} = \begin{bmatrix} 0 & C_{P1} & 0 \end{bmatrix}, \quad
\tilde{F} = \begin{bmatrix} F_R & 0 & F_P \end{bmatrix}.
\end{cases}
\tag{10.23}
$$

The step response [Figure 10.7(b)] shows that it is almost the same as that of the servo system directly using the state feedback.

Program 10.4   Servo system

```
1  % Plant
2  ap=[0 0 1 0; 0 0 0 1; 0 -2 -6 0.002; 0 40 6 -0.04];
3  bp=[0; 0; 16; -15]; cp1=[1 0 0 0];
4
5  na=length(ap);
6  Co=ctrb(ap,bp); unco=length(ap)-rank(Co)
7  Ob=obsv(ap,cp1); unob = length(ap)-rank(Ob)
8
9  % Internal model
10 AR=0; BR=1;
11
12 % A_bar, B_bar, C_bar
13 A=[AR -BR*cp1; zeros(na,1) ap];
14 B=[0; bp]; Bi=[BR; zeros(na,1)];
15 C=[0 cp1];
16
17 % Feedback gain
18 RF=1; QF=diag([1 10 10 1 1]);
19 F=lqr(A,B,QF,RF); FR=F(1,1); Fp=F(1,2:na+1);
20
21 % Internal model + state feedback
22 sys=ss(A-B*F,Bi,C,0); sysu=ss(A-B*F,Bi,-F,0);
23 figure(1), hold on;
24 step(sys); grid;
25 figure(2), hold on;
26 step(sysu); grid;
27
28 % Observer gain
29 po=[-10+2i -10-2i -20+5i -20-5i];
30 L=(place(ap',cp1',po))'; eig(ap-L*cp1)
31
```

```
32  % A_tilde, B_tilde, C_tilde
33  At=[AR -BR*cp1 zeros(1,na); zeros(na,1) ap zeros(na,na);
        zeros(na,1) L*cp1 ap-L*cp1];
34  Bt=[0; bp; bp];
35  Bit=[BR; zeros(2*na,1)];
36  Ct=[0 cp1 zeros(1,na)];
37  Ft=[FR zeros(1,na), Fp];
38
39  % Internal model + state observer + state feedback
40  sysob=ss(At-Bt*Ft,Bit,Ct,0); sysobu=ss(At-Bt*Ft,Bit,-Ft,0);
41  figure(1), step(sysob)
42  figure(2), step(sysobu);
```

──────────── **Problems** ────────────

⟨ **Basic Level** ⟩

[1] A potentiometer in a control system is used as
    A. an error detector   B. an actuator
    C. an amplifier   D. a feedback element.

[2] A tachogenerator in a control system is used as
    A. an error detector   B. an actuator
    C. an amplifier   D. a feedback element.

[3] If the closed-loop system in **Figure 10.8** operates with a damping ratio of
0.717 for a pair of dominant poles, find the gain K and the location of all
closed-loop poles and zeros.

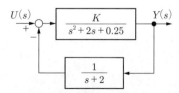

Figure 10.8 Problem 3

[4] Consider a continuous-time system $\dot{x}(t) = x(t) + u(t)$. Use $Q = 2$ and

$R = 1/2$ to optimize the LQR performance index (8.23). Solve the algebraic Riccati equation (8.25) and find the feedback gain F (8.24). Find the closed-loop system matrix $A - BF$ and verify that it is stable.

[5] The internal model of a step signal is $1/s$. Insert the model into a control system to construct a servo system [Figure 10.6(a)]. Design a state-feedback control law that yields a step response with $M_P \leq 10\%$, $T_r \leq 0.1$ s, and $T_s \leq 0.5$ s for a plant

$$\dot{x}(t) = \begin{bmatrix} 0 & 1 \\ -7 & -9 \end{bmatrix} x(t) + \begin{bmatrix} 0 \\ 1 \end{bmatrix} u(t) \text{ and } y(t) = \begin{bmatrix} 4 & 1 \end{bmatrix} x(t).$$

[6] Consider the problem of designing a PID controller $C(s) = K_P[1 + 1/(T_i s) + T_d s]$ for an arm robot $P(s) = \beta/[s(s + \alpha)]$. The parameters take the values $\alpha \in [\alpha_m, \ \alpha_M]$ and $\beta \in [\beta_m, \ \beta_M]$ ($\alpha_m = 4.17$, $\alpha_M = 8.33$, $\beta_m = 22.1$, $\beta_M = 44.2$). Choose the nominal values $\bar{\alpha} = (\alpha_M + \alpha_m)/2 = 6.25$ and $\bar{\beta} = (\beta_M + \beta_m)/2 = 33.2$ to build a nominal plant $\bar{P}(s) = \bar{\beta}/[s(s + \bar{\alpha})]$.

    A.   Choose the closed-loop poles to be $\{-20 \pm j10, \ -50\}$. Find PID parameters for the nominal plant $\bar{P}(s)$.

    B.   Check if Condition (10.6) holds for all possible combinations of α and β [check the combinations of α (α_m and α_M) and β (β_m and β_M)].

    C.   Choose five values of α from $[\alpha_m, \ \alpha_M]$ and five values of β from $[\beta_m, \ \beta_M]$ (including the boundaries), draw Nyquist plots for those combinations, and verify the decision given in B.

    D.   Draw Bode plots for the combinations of the parameters in C and verify the decision given in B.

    E.   Draw Bode plots for the sensitivity functions of the closed-loop control system $S(s) = 1/[1 + P(s)C(s)]$ for the combinations of the parameters in C and explain the tracking and disturbance-rejection characteristics for a step signal and for a sine wave $\sin t$ (Assume that a disturbance is added to the output).

    F.   Use Simulink to build a simulator and validate the conclusions given in E.

[7] Consider the two-mass system in Figure 9.11. Let the parameters be

$$A = \begin{bmatrix} -31.31 & 0 & -2.833 \times 10^4 \\ 0 & -10.25 & 8\,001 \\ 1 & -1 & 0 \end{bmatrix},$$

$$B = \begin{bmatrix} 28.06 \\ 0 \\ 0 \end{bmatrix}, \quad B_d = \begin{bmatrix} 0 \\ 7.210 \\ 0 \end{bmatrix}, \quad C = \begin{bmatrix} 1 & 0 & 0 \end{bmatrix}. \tag{10.24}$$

Solve the following problems:

A. Check the controllability and observability of the plant.

B. Add an internal model of a step signal to the system and build an augmented system.

C. Check the controllability and observability of the augmented system.

D. Use the LQR method ($Q = \text{diag}\{10^6, 1, 1\}$ and $R = 1$) to design a state observer.

E. Use the LQR method ($Q = \text{diag}\{1, 1, 1, 10\}$ and $R = 1$) to design state feedback.

F. Use Simulink to build a simulator and verify the time response of the system for $r(t) = 1\,000 \times 1(t)$ and $d(t) = 5 \times 1(t)$.

Advanced Level

[1] Consider the system in **Figure 10.9**, in which both the input and the disturbance are step signals. Solve the following problems:

A. Find the error of the system, $E(s) = G_{eu}(s)U(s) + G_{ed}(s)D(s)$.

B. Find the steady-state error, e_{ss}.

C. Find the sensitivity of e_{ss} to K_1 and K_2, $\partial e_{ss}/\partial K_1$ and $\partial e_{ss}/\partial K_2$, at $K_1 = 100$ and $K_2 = 0.1$.

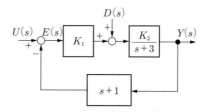

Figure 10.9   Advanced problem 1

[2] Consider a tracking problem for a plant $P(s) = 50/[s(s + 10)]$. Let the reference input be the unit step signal. Find an internal model of the input, insert the model in the control system and solve the following design problems:

A. Choose weighting matrices Q_K and R_K and design the optimal control law $\bar{F}$ in (10.19) that optimizes (10.21). Compare the difference for a different choice of Q_K and R_K based on the time response to the reference input.

B. Design a full-order state observer to carry out state-feedback control. Choose the real parts of the poles for the observer to be 0.5, 1.0, 2.0, 5.0, and 10.0 of that of the dominant poles of the closed-loop control system and investigate the effect of the observer on reference tracking through simulations.

[3] A speed-control system for an electric vehicle contains two control loops: an inner current-control loop and an outer speed-control loop. Let

$$G_1(s) = \frac{80}{0.0033s + 1}, \ G_2(s) = \frac{200}{0.2s + 1}, \ G_B(s) = \frac{0.0025}{0.0018s + 1},$$

$$G_F(s) = \frac{1}{T_F s + 1}, \text{ and } G_C(s) = K_P\left(1 + \frac{1}{T_i s}\right)$$

in the inner current-control loop. Consider the problem of designing the inner current-PI-controller (**Figure 10.10**).

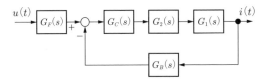

Figure 10.10 Advanced problem 3

A. Convert the system to a unity-feedback system.

B. Find T_F in $G_F(s)$ to make the feedforward term, $G_F(s)/G_B(s)$ be a gain term for the unity-feedback system.

C. Find the transfer function of the plant for the PI design.

D. Use (9.9) to find reduced-order systems for the plant.

E. Let $M_p \leq 5\%$, $T_r \leq 1.5\,\text{s}$, and $T_s \leq 2\,\text{s}$. Find the parameters of the PI controller $G_C(s)$ for different reduced-order systems.

F. Use MATLAB and/or Simulink to verify the designed results and compare the treatments of small time constants.

References

1) 中野 道雄, 美多 勉：制御基礎理論—古典から現代まで—, コロナ社 (2014)

2) N. S. Nise：Control System Engineering, 8th Ed., Wiley (2019)

3) A. Gilat：MATLAB: An Introduction with Applications, 6th Ed., Wiley (2016)

4) S. L. Campbell, J.-P. Chancelier, and R. Nikoukhah：Modeling and Simulation in Scilab/Scicos with ScicosLab 4.4, 2th Ed., Springer (2009)

5) A. Vande Wouwer, P. Saucez, and C. Vilas Fernández：Simulation of ODE/PDE Models with MATLAB$^{\circledR}$, OCTAVE and SCILAB, Springer (2014)

6) G. F. Franklin, J. D. Powell, and A. Emami-Naeini：Feedback Control of Dynamic Systems, 7th Ed., Pearson (2015)

7) 細江 繁幸：システムと制御, オーム社 (2003)

8) 斉藤 制海, 徐 粒：制御工学—フィードバック制御の考え方—, 森北出版 (2016)

9) J. C. Doyle, B. A. Francis, and A. R. Tannenbaum：Feedback Control Theory, Macmillan Publishing Company (1992)

10) B. Shahian and M. Hassul：Control System Design Using MATLAB, Prentice Hall (1993)

11) 前田 肇, 杉江 俊治：アドバンスト制御のためのシステム制御理論, 朝倉書店 (1990)

12) H. Taguchi and M. Araki：Two-degree-of-freedom PID controllers—Their functions and optimal tuning—, Proc. IFAC Digital Control: Past, Present and Future of PID Control, pp. 91–96, Terrassa, Spain (2000)

13) S. Skogestad and I. Postlethwaite：Multivariable Feedback Control —Analysis and Design—, 2nd Ed., John Willy & Sons, Ltd (2005)

14) R. Y. Chiang：Modern Robust Control Theory, Ph.D. Dissertation, Univ. Southern California (1988)

15) K. J. Åström and B. Wittenmark：Compter-Controlled Systems—Theory and Design—, 3rd Ed., Prentice Hall (1997)

16) 森 泰親：演習で学ぶ PID 制御, 森北出版 (2011)

17) P. Dorato, C. Abdallah, and V. Cerone：Linear-Quadratic Control–An Introduction–, Prentice Hall (1995)

18) 北森 俊行 ほか：最適な制御系設計法と各種制御方式の基礎・理論・応用の実際, アイ・エヌ・ジー出版部 (1993)
19) 岩井 善太, 井上 昭, 川路 茂保：オブザーバ, コロナ社 (1988)
20) R. V. Dukkipati：Analysis and Design of Control Systems Using MATLAB, New Age International Publishers (2006)
21) 平田 光男：Arduino と MATLAB で制御系設計を始めよう！, TechShare (2020)
22) 胡 寿松, 薛 安克：自動控制原理題海大全, 科学出版社 (2010)
23) 翁 貽方：自動控制理論例題習題集・考研試題解析, 機械工業出版社 (2006)
24) R. Cowell, 佘 錦華：マスターしておきたい 技術英語の基本—決定版—, コロナ社 (2015)
25) 川田 昌克 編著：倒立振子で学ぶ制御工学, 森北出版 (2018)
26) 日本機械学会：機械工学便覧 α. 基礎編, 日本機械学会 (2007)

Answers to Problems

★ Chapter 1

── Basic Level ──

【1】

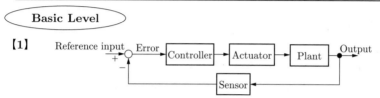

Figure A1.1 Structure of control system

【2】 A.

【3】 A, C, and D.

【4】 A.

【5】 C.

【6】 Sequential control: C, D, and E;　　Feedback control: A and B.

【7】 A, B, and C.

【8】

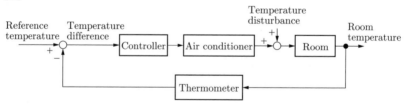

Figure A1.2 Room-temperature-control system

【9】

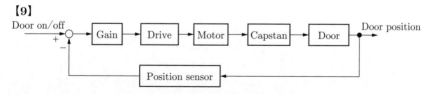

Figure A1.3 Door-control system

Advanced Level

[1] (1) Turn on a power switch.   (2) Cook over low heat at the beginning.   (3) Cook over high heat at the middle stage.   (4) Weaken the heat when the water starts boiling.   (5) Turn off the power switch and steam boiled rice.

[2] (1) Big driving force ensuring quick response.   (2) Short response time.   (3) High enough positioning precision.   (4) Robust with regard to temperature and environment changes.   (5) Small vibrations. (6) Small enough volume.

★ Chapter 2

Basic Level

[1]

Table A2.1

No.	Cartesian coordinates	Polar coordinates	Conjugate numbers
A	$\left(\dfrac{3}{13}, -\dfrac{2}{13}\right)$	$\dfrac{\sqrt{13}}{13}e^{j\theta}\left(\theta = -\tan^{-1}\dfrac{2}{3}\right)$	$\dfrac{3 + j2}{13}$
B	$\left(-\dfrac{7}{2}, -13\right)$	$\dfrac{5\sqrt{29}}{2}e^{j\theta}\left(\theta = -\pi + \tan^{-1}\dfrac{26}{7}\right)$	$-\dfrac{7}{2} + j13$
C	$(1, \sqrt{3})$	$2e^{j\theta}\ (\theta = 60°)$	$1 - j\sqrt{3}$
D	$(1, -1)$	$\sqrt{2}e^{j\theta}\ (\theta = -45°)$	$1 + j$

[2] $|z_1| = 2$ and $\angle z_1 = 30°$, $|z_2| = 3$ and $\angle z_2 = 60°$.

[3] A. $\dfrac{3 + 2\sqrt{3}}{2} + j\dfrac{2 + 3\sqrt{3}}{2}$ B. $\dfrac{2\sqrt{3} - 3}{2} + j\dfrac{2 - 3\sqrt{3}}{2}$ C. $j6$

　　D. $\dfrac{\sqrt{3} - j}{3}$.

[4] A. $0.52\,\text{rad}$ B. $0.79\,\text{rad}$ C. $1.05\,\text{rad}$

　　D. $114.59°$ E. $85.94°$ F. $28.65°$.

[5] A. $\dfrac{\sqrt{3}}{2}$ B. $\dfrac{\sqrt{3}}{3}$ C. $\dfrac{\sqrt{2}}{2}$ D. $-\dfrac{\sqrt{3}}{6}$.

[6] Since $\cos^2\theta = \dfrac{1 + \cos 2\theta}{2}$, $f(t) = (10\cos t)^2 = 100 \times \dfrac{1 + \cos 2t}{2} = 50 + 50\cos 2t$. 50 is a periodic function with its period being any positive number and the period of $\cos 2t$ is π s. Thus, the period of $f(t)$ is π s.

[7] A. $F(s) = \dfrac{6}{s} + \dfrac{4}{s+1}$

B. $F(s) = 1 - \dfrac{a}{s+a} = \dfrac{s}{s+a}$

C. $F(s) = \dfrac{\omega}{(s-a)^2 + \omega^2}$

D. $F(s) = \dfrac{6}{(s+5)^4}$.

[8] A. $f(t) = \cos\omega(t-3)$

B. $f(t) = 2e^{-2t} + e^{-t}$

C. $f(t) = e^{-2t} - e^{-5t}$

D. $f(t) = \dfrac{15e^{-4t} + 17e^{4t}}{8}$.

[9] A. $\mathcal{L}[f(t)] = \displaystyle\int_0^{+\infty} e^{-bt}\sin at \times e^{-st}\mathrm{d}t = \int_0^{+\infty} \sin at \times e^{-\tilde{s}t}\mathrm{d}t \ (\tilde{s} := s+b)$.

Thus, $\mathcal{L}[f(t)] = \dfrac{a}{\tilde{s}^2 + a^2} = \dfrac{a}{(s+b)^2 + a^2}$.

```
syms a b t; f=exp(-b*t)*sin(a*t), F=laplace(f)
```

B. Using $\sin^2\theta = \dfrac{1 - \cos 2\theta}{2}$ gives $\mathcal{L}[f(t)] = \dfrac{2\omega^2}{s(s^2 + 4\omega^2)}$.

```
syms w t; f=sin(w*t)*sin(w*t), F=laplace(f)
```

[10] $\displaystyle\int_0^1 (1-t)e^{-st}\mathrm{d}t = \int_0^1 e^{-st}\mathrm{d}t - \int_0^1 te^{-st}\mathrm{d}t$

$= \left[-\dfrac{e^{-st}}{s}\right]_0^1 - \left(\left[-\dfrac{t\cdot e^{-st}}{s}\right]_0^1 - \int_0^1 \left(-\dfrac{e^{-st}}{s}\right)\mathrm{d}t\right)$

$= \dfrac{e^{-s}}{s^2} - \dfrac{1}{s^2} + \dfrac{1}{s}$.

[11] A. $F(s) = \dfrac{1}{(s+1)(s-2)(s+3)} = -\dfrac{1}{6}\dfrac{1}{s+1} + \dfrac{1}{15}\dfrac{1}{s-2} + \dfrac{1}{10}\dfrac{1}{s+3}$.

Thus, $f(t) = -\dfrac{1}{6}e^{-t} + \dfrac{1}{15}e^{2t} + \dfrac{1}{10}e^{-3t}$.

```
syms s w; F=1/((s+1)*(s-2)*(s+3)), f=ilaplace(F)
```

B. $F(s) = \dfrac{a}{\omega}\sin\omega t$.

```
syms w s a; F=a/(s^2+w^2), f=ilaplace(F)
```

[12] C.

Advanced Level

[1] $\mathcal{L}[f(t)] = \displaystyle\int_0^{+\infty} f(t)e^{-st}\mathrm{d}t = \sum_{k=0}^{+\infty} \int_{2bk}^{2b(k+1)} f(t)e^{-st}\mathrm{d}t$. Letting $t = \tau + 2bk$

yields $\displaystyle\int_{2bk}^{2b(k+1)} f(t)e^{-st}\mathrm{d}t = e^{-s2bk}\int_0^{2b} f(\tau)e^{-s\tau}\mathrm{d}\tau$. $\displaystyle\int_0^{2b} f(\tau)e^{-s\tau}\mathrm{d}\tau =$

$\dfrac{1}{s^2}\left(1 - e^{-bs}\right)^2$. Since $\left|e^{-s2b}\right| = e^{-\mathrm{Re}(s)2b} < 1$ for $\mathrm{Re}(s) > 0$, $\displaystyle\sum_{k=0}^{\infty} e^{-s2bk} =$

$\dfrac{1}{1 - e^{-s2b}}$ and $\mathcal{L}[f(t)] = \dfrac{1}{s^2}\dfrac{1 - e^{-sb}}{1 + e^{-sb}}$.

[2] $F(s) = \dfrac{-1}{s} + \dfrac{1}{s^2} + \dfrac{1}{s+1}$. Thus, $f(t) = -1 + t + e^{-t}$.

```
syms s; F=1/(s^2*(s+1)), f=ilaplace(F)
```

★ Chapter 3

Basic Level

[1] (a) $x(t) = \begin{bmatrix} x_1(t) \\ x_2(t) \end{bmatrix} = \begin{bmatrix} y(t) \\ \dot{y}(t) \end{bmatrix}$, $\dot{x}_1(t) = x_2(t)$,

$\dot{x}_2(t) = -\dfrac{b_1 + b_2}{M}x_2(t) + \dfrac{b_2}{M}\dot{r}(t)$,

$y(t) = x_1(t)$, $\dot{x}(t) = Ax(t) + Bu(t)$, $y(t) = Cx(t)$,

$A = \begin{bmatrix} 0 & 1 \\ 0 & -\dfrac{b_1 + b_2}{M} \end{bmatrix}$, $B = \begin{bmatrix} 0 \\ \dfrac{b_2}{M} \end{bmatrix}$, $C = \begin{bmatrix} 1 & 0 \end{bmatrix}$.

(b) $x(t) = \begin{bmatrix} y(t) \\ \dot{y}(t) \end{bmatrix}$, $\dot{x}(t) = Ax(t) + Bu(t)$, $y(t) = Cx(t)$,

$A = \begin{bmatrix} 0 & 1 \\ 0 & -\dfrac{b}{M} \end{bmatrix}$, $B = \begin{bmatrix} 0 \\ \dfrac{1}{M} \end{bmatrix}$, $C = \begin{bmatrix} 1 & 0 \end{bmatrix}$.

(c) $x(t) = \begin{bmatrix} y(t) \\ \dot{y}(t) \end{bmatrix}$, $\dot{x}(t) = Ax(t) + Bu(t)$, $y(t) = Cx(t)$,

$A = \begin{bmatrix} 0 & 1 \\ -\dfrac{K}{M} & -\dfrac{b}{M} \end{bmatrix}$, $B = \begin{bmatrix} 0 \\ \dfrac{1}{M} \end{bmatrix}$, $C = \begin{bmatrix} 1 & 0 \end{bmatrix}$.

[2] (a) $\dot{i}(t) = Ai(t) + Bv_i(t)$, $v_o(t) = Ci(t) + Dv_i(t)$,

$A = -\dfrac{R}{L}$, $B = \dfrac{1}{L}$, $C = -R$, $D = 1$.

(b) $x(t) = \begin{bmatrix} v_0(t) \\ \dot{v}_0(t) \end{bmatrix}$, $\dot{x}(t) = Ax(t) + Bu(t)$, $x(t) = Cx(t)$,

$$A = \begin{bmatrix} 0 & 1 \\ -\dfrac{1}{LC} & -\dfrac{R}{L} \end{bmatrix}, B = \begin{bmatrix} 0 \\ \dfrac{1}{LC} \end{bmatrix}, C = \begin{bmatrix} 1 & 0 \end{bmatrix}, D = 0.$$

[3] A.   [4] C.   [5] B.

[6] (a) $M\dfrac{d^2y(t)}{dt^2} + (b_1 + b_2)\dfrac{dy(t)}{dt} = b_2\dfrac{dr(t)}{dt}$. $G(s) = \dfrac{b_2 s}{Ms^2 + (b_1 + b_2)s}$.

   (b) $M\dfrac{d^2y(t)}{dt^2} = K[x(t) - y(t)] - b\dfrac{dy(t)}{dt}$ and $K[x(t) - y(t)] = f(t)$. Thus,

   $M\dfrac{d^2y(t)}{dt^2} + b\dfrac{dy(t)}{dt} = f(t)$. $G(s) = \dfrac{1}{Ms^2 + bs}$.

   (c) $M\dfrac{d^2y(t)}{dt^2} + b\dfrac{dy(t)}{dt} + Ky(t) = f(t)$. $G(s) = \dfrac{1}{Ms^2 + bs + K}$.

[7] (a) $\dfrac{v_i(t) - v_o(t)}{R} = i(t)$ and $v_o(t) = L\dfrac{di(t)}{dt}$. Thus, $L\dfrac{dv_o(t)}{dt} + Rv_o(t) =$

   $L\dfrac{dv_i(t)}{dt}$. $G(s) = \dfrac{V_o(s)}{V_i(s)} = \dfrac{Ls}{R + Ls}$.

   (b) $v_o(t) = v_i(t) - L\dfrac{di(t)}{dt} - v_C(t)$, $i(t) = C\dfrac{dv_C(t)}{dt}$, and $v_o(t) = Ri(t)$.

   Thus, $LC\dfrac{d^2v_o(t)}{dt^2} + RC\dfrac{dv_o(t)}{dt} + v_o(t) = RC\dfrac{dv_i(t)}{dt}$. $G(s) = \dfrac{V_o(s)}{V_i(s)} =$

   $\dfrac{RCs}{LCs^2 + RCs + 1}$.

[8] $(s^2 + 3s + 2)Y(s) = sU(s)$. Thus, $\ddot{y}(t) + 3\dot{y}(t) + 2y(t) = \dot{u}(t)$.

[9] (a) $f(t) = K_2[y(t) - x(t)]$ and $K_2[y(t) - x(t)] = K_1 x(t)$. $G(s) = \dfrac{K_1 + K_2}{K_1 K_2}$.

   (b) $f(t) = b_2\dfrac{d[y(t) - x(t)]}{dt}$ and $b_2\dfrac{d[y(t) - x(t)]}{dt} = b_1\dfrac{dx(t)}{dt}$.

   $G(s) = \dfrac{b_1 + b_2}{b_1 b_2}\dfrac{1}{s}$.

[10] $G(s) = \dfrac{1}{2s^2 + s + 2}$.

[11] $\dfrac{p_{s0}}{1 + wx^2} = \dfrac{p_{s0}}{1 + wx_0^2} - \dfrac{2p_{s0}wx_0}{(1 + wx_0^2)^2}[x(t) - x_0]$. Letting $K = -\dfrac{2p_{s0}wx_0}{(1 + wx_0^2)^2}$

   yields $\Delta p_1(t) = K\Delta x(t)$.

[12] Since $\sqrt{h(t)} = \sqrt{h_0} + \dfrac{1}{2\sqrt{h_0}}[h(t) - h_0]$,

   $\dfrac{d\Delta h(t)}{dt} = -\dfrac{c}{2S\sqrt{h_0}}\Delta h(t) + \dfrac{1}{S}\Delta q_i(t)$. $G(s) = \dfrac{\Delta H(s)}{\Delta Q_i(s)} = \dfrac{1/S}{s + c/(2S\sqrt{h_0})}$.

Advanced Level

[1] D.

[2] (a) $x(t) = \begin{bmatrix} x_1(t) \\ \dot{x}_1(t) \\ x_2(t) \\ \dot{x}_2(t) \end{bmatrix}$. $A = \begin{bmatrix} 0 & 1 & 0 & 0 \\ -\dfrac{K}{M_1} & -\dfrac{b}{M_1} & \dfrac{K}{M_1} & \dfrac{b}{M_1} \\ 0 & 0 & 0 & 1 \\ \dfrac{K}{M_2} & \dfrac{b}{M_2} & -\dfrac{K}{M_2} & -\dfrac{b}{M_2} \end{bmatrix}$, $B = \begin{bmatrix} 0 \\ \dfrac{1}{M_1} \\ 0 \\ 0 \end{bmatrix}$,

and $C = [0\ 0\ 1\ 0]$.

(b) $x(t) = \begin{bmatrix} x_1(t) \\ \dot{x}_1(t) \\ x_2(t) \\ \dot{x}_2(t) \end{bmatrix}$. $A = \begin{bmatrix} 0 & 1 & 0 & 0 \\ -\dfrac{K_1+K_2}{M_1} & -\dfrac{b}{M_1} & \dfrac{K_2}{M_1} & \dfrac{b}{M_1} \\ 0 & 0 & 0 & 1 \\ \dfrac{K_2}{M_2} & \dfrac{b}{M_2} & -\dfrac{K_2}{M_2} & -\dfrac{b}{M_2} \end{bmatrix}$,

$B = \begin{bmatrix} 0 \\ 0 \\ 0 \\ \dfrac{1}{M_2} \end{bmatrix}$, and $C = [0\ 0\ 1\ 0]$.

[3] $i_{R1}(t) = i_{C1}(t) + i_{R2}(t)$ and $i_{R2}(t) = i_{C2}(t)$ yield

$$\frac{V_o(s)}{V_i(s)} = \frac{R_1 R_2 C_1 C_2 s^2 + (R_1 + R_2)C_2 s + 1}{R_1 R_2 C_1 C_2 s^2 + [(R_1 + R_2)C_2 + R_1 C_1]s + 1}.$$

[4] [5] Omitted.

★ Chapter 4

Basic Level

[1] (a) $G_{yr}(s) = \dfrac{G_1(s)G_2(s)}{1 + G_2(s)}$ (b) $G_{yr}(s) = \dfrac{G_1(s)G_2(s)}{1 + G_1(s)}$

(c) $G_{yr}(s) = \dfrac{[G_2(s) + G_1(s)G_3(s)]G_4(s)}{1 + G_3(s)G_4(s)}$

(d) $G_{yr}(s) = \dfrac{G_1(s)G_2(s)}{1 + G_1(s)G_2(s)G_3(s)}$

(e) $G_{yr}(s) = \dfrac{G_1(s)G_2(s)}{1 + G_1(s)H_1(s) - G_1(s)G_2(s)H_1(s)H_2(s)}$

(f) $G_{yr}(s) = \dfrac{G_1(s)G_2(s)G_3(s)}{1 + G_2(s)G_3(s) - G_1(s)G_2(s)}$.

[2] $\begin{bmatrix} Y_1(s) \\ Y_2(s) \end{bmatrix} = \begin{bmatrix} \dfrac{G_1(s)}{1 + G_1(s)G_2(s)G_3(s)G_4(s)} & \dfrac{-G_1(s)G_3(s)G_4(s)}{1 + G_1(s)G_2(s)G_3(s)G_4(s)} \\ \dfrac{G_1(s)G_2(s)G_3(s)}{1 + G_1(s)G_2(s)G_3(s)G_4(s)} & \dfrac{G_3(s)}{1 + G_1(s)G_2(s)G_3(s)G_4(s)} \end{bmatrix} \begin{bmatrix} R_1(s) \\ R_2(s) \end{bmatrix}.$

[3] Example commands:

```
s=tf('s'); G0=10/(s*(s+1)),
G01=(10*exp(-0.1*s))/(s*(s+1)),
G3=(10*exp(-3*s))/(s*(s+1)), nyquistplot(G0,G01,G3);
```

[4]
```
s=tf('s'); bode(G0,G01,G3); grid on;
```

[5] A.
```
num=[1], den=[1 1 0], G=tf(num, den), nyquistplot(G)
```

B.~F.   Omitted.

[6] A.
```
bode(G); grid on;
```

B.~F.   Omitted.

[7] A.
```
num=[10], den=conv([1 1],[10 1]), G=tf(num, den), bode(G);
```

B.~F.   Omitted.

Advanced Level

[1] (a) $G(s) = \dfrac{[G_1(s) + G_4(s)]G_2(s)}{1 + G_2(s)G_3(s) + G_1(s)G_2(s)}$

(b) $G(s) = \dfrac{G_1(s)G_2(s)G_3(s)G_4(s)}{1 + G_3(s)G_4(s) + G_2(s)G_3(s) + G_1(s)G_2(s)}$

(c) $G(s) = \dfrac{G_1(s)G_2(s)G_3(s)}{1 + G_1(s)G_2(s)H_1(s) + G_2(s)G_3(s)H_2(s) + G_1(s)G_2(s)G_3(s)H_3(s)}.$

[2] A.   (a) $\dfrac{3.16(s+1)}{s^2(0.2s+1)}$ (b) $\dfrac{10}{(0.5s+1)(0.05s+1)}.$

B.   Omitted.

[3] Similarity: gains,   Difference: phases.

★ Chapter 5

Basic Level

[1] A.

[2] A. $G(s) = \dfrac{C(s)P(s)}{1 + C(s)P(s)} = \dfrac{1}{\dfrac{1}{K}s + 1}.$ $\dfrac{1}{K} = 0.05.$ $K = 20.$

B. $G(s) = \dfrac{C(s)P(s)}{1 + C(s)P(s)} = \dfrac{\dfrac{50}{10 + 50K}}{\dfrac{1}{10 + 50K}s + 1} \cdot \dfrac{1}{10 + 50K} = 0.05.$

$K = 0.2.$

[3] A. $G(s) = \dfrac{C(s)P(s)}{1 + C(s)P(s)} = \dfrac{\dfrac{K}{9}(Ts + 1)}{s^2 + \dfrac{KT}{9}s + \dfrac{K}{9}} \cdot 2\zeta\omega = \dfrac{KT}{9} = \dfrac{2}{3},$

$\omega^2 = \dfrac{K}{9} = \dfrac{2}{9}. \quad \therefore \omega = \dfrac{\sqrt{2}}{3}, \zeta = \dfrac{\sqrt{2}}{2} = 0.707.$

B. $KT^2 = 18.$

[4] B.

[5] A. No solution   B. $K > -a$ C. $-a < K < 0$ D. $K > 0.$

[6] A. $-1, -4, -5$ (Stable).

B. $1, 3, -\dfrac{2}{3}, 20, 6$ (Unstable, two unstable poles).

C. If $a_0, a_1, a_2, a_3 > 0$ or $a_0, a_1, a_2, a_3 < 0$ and $a_1 a_2 > a_0 a_3$, system is stable.

D. $1, 2, -\dfrac{1}{2}, 9, \dfrac{16}{9}, 5$ (Unstable, two unstable poles).

E. $1, 8, 9, \dfrac{64}{9}, 1$ (Stable).

F. $1, 1, 2, -6, 2$ (Unstable, two unstable poles).

[7] A. $0 < K < 12$ B. $K > 25$ C. $K > \dfrac{20}{3}.$

[8] A. $-1 < K < 1$ B. $0 < K < -5 + 3\sqrt{5}.$

[9] B.

[10] $T = 0$ and $T = 2$: Stable, $T = 4$: Unstable.

[11] B.

[12] A. $Z = 2$ (Unstable)   B. $Z = 2$ (Unstable)   C. $Z = 0$ (Stable)
D. $Z = 0$ (Stable)   E. $Z = 0$ (Stable)   F. $Z = 2$ (Unstable)
G. $Z = 0$ (Stable)   H. $Z = 2$ (Unstable)   I. $Z = 1$ (Unstable).

[13] C. gm $= \infty$ and PM $= 65.5°$.   D. gm $= 3.52$ dB and PM $= 11.4°$.
E. gm $= 25.9$ dB and PM $= 180°$.   F. gm $= 19.8$ dB and PM $= -180°$.
G. gm $= -6.02$ dB and PM $= 60°$.

[14] Stable. gm $= 9.54$ dB and PM $= 179°$.

```
num=[1], den=[1 2 2 1], G=tf(num, den);
bode(G); margin(G), grid on;
```

〖15〗 A. Bode plots for $K = 1$ show that gm $= 55\,$dB. $K = 55 - 10 = 45\,$dB (or $K = 177.8$).

   B. plots for $K = 1$ show that gm $= 76.5\,$dB. $K = 76.5 - 10 = 66.5\,$dB (or $K = 2113$).

Advanced Level

〖1〗 The first column of Routh's table is $\left(1, 3, \dfrac{3\epsilon - 3}{\epsilon}, 1, 0\right)$. $\dfrac{3\epsilon - 3}{\epsilon} < 0$ for $\epsilon = 0^+$. Unstable (two unstable poles).

Table A5.1

1st row (s^4)	1	1	1	+
2nd row (s^3)	3	3		+
3rd row (s^2)	ϵ	1		$+\ (\epsilon > 0)$
4th row (s^1)	$\dfrac{3\epsilon - 3}{\epsilon}$			−
5th row (s^0)	1			+

〖2〗 Omitted.

〖3〗 C (GM > 1 for many systems, for example, the one in basic-level problem 12 E in Chapter 5. However, some systems may have GM < 1. For example, in Advanced problem 2 in Chapter 5.

〖4〗 A. $-48 < K < 480$.   B.~D. Omitted.

〖5〗 $K_1 > 0$ and $K_2 > K_1/4 + 4/K_1$.

〖6〗 A. Cannot.

   B. $1 - 2s : (-1 < K < 5)$ $\dfrac{1}{1 + 2s} : (-1 < K)$ $\dfrac{1 - s}{1 + s} : (-1 < K < 11)$

   C. It is hard to use Padé approximation to find a precise stability condition.

〖7〗 A. Stable   B. Stable   C. $C(s) = -0.5$ is stable $(-1 < K < 0)$

   D. $C(s) = -1.5$ is stable $\left(-\dfrac{18}{11} < K < -1\right)$.

★ Chapter 6

Basic Level

〖1〗 $G_{yd}(s) = \dfrac{Y(s)}{D(s)} = \dfrac{P(s)}{1 + C(s)P(s)} = \dfrac{50s}{s^3 + 90s^2 + 2\,500s + 25\,000}$. $g(1) = -54\,$dB, that is, $|G_d(j1)| = 0.002$. Disturbance $\sin t$ is reduced to $\dfrac{1}{500}$.

[2] Step 1: $T(s) = \dfrac{P(s)C(s)}{1 + P(s)C(s)} = \dfrac{80s^2 + 2\,500s + 25\,000}{s^3 + 90s^2 + 2\,500s + 25\,000}$.

Steps 2~3: Omitted.   Step 4: $L = 0.0008\,\text{s}$.

[3] $2\zeta\omega_n = 6$ and $\omega_n^2 = 36$. Thus, $\omega_n = 6$. $\zeta = 0.5$.

[4] A.

[5] B.

[6] A.

[7]
```
n=[20], d=[1 4 25], G=tf(n,d), t=0:0.01:5; lsim(G,t,t);
xlabel('time'); ylabel('r(t) and y(t)');
title('Ramp response'); grid;
```

$e_{ss} = \infty$.

[8] A. Characteristic equation is $s^2 + 10s + 100 = 0$. $\omega_n = \sqrt{100} = 10\,\text{rad/s}$. $2\zeta\omega_n = 10$ yields $\zeta = 0.5$.

B. $T_r = 0.164\,\text{s}$. $M_P\% = 16.3\%$. $T_s = 0.529\,\text{s}$.

C. Unit step input: $e_{ss} = 0$. Unit ramp input: $e_{ss} = 0.1$. Unit parabolic input: $e_{ss} = \infty$.

[9] A. Type 0.   B. $K_p = 127$. Unit step: $e_{ss} = 7.8 \times 10^{-3}$. Unit ramp and parabolic: $e_{ss} = \infty$.

Advanced Level

[1] gm $= \infty$ and PM changes from $70.1°$ to $59.8°$. Since $G_{yd0}(j1) = G_{yd}(j1) = -12.7\,\text{dB}$, the disturbance $\sin t$ is reduced to 0.23 times. Since $G_{yd0}(j1) = -11.3\,\text{dB}$ and $G_{yd}(j1) = -8.89\,\text{dB}$, the disturbance $\sin 5t$ is reduced to about 0.3 times. The changes in the parametes deteriorates the performance of disturbance rejection by $2.41\,\text{dB}$ (0.09 times).

[2] $\sin(\omega_d t + \phi) = 0$ for $\omega_d t + \phi = k\pi$ $(k = 0, 1, 2, \cdots)$. Thus, $e(t) \geq 0$ for $0 \leq t \leq \dfrac{\pi - \phi}{\omega_d}$, $e(t) \leq 0$ for $\dfrac{(2k+1)\pi - \phi}{\omega_d} \leq t \leq \dfrac{(2k+2)\pi - \phi}{\omega_d}$, and $e(t) \geq 0$ for $\dfrac{(2k+2)\pi - \phi}{\omega_d} \leq t \leq \dfrac{(2k+3)\pi - \phi}{\omega_d}$ $(k = 0, 1, 2, \cdots)$. $\displaystyle\int_0^\infty |e(t)| \mathrm{d}t =$

$\displaystyle\int_0^{(\pi-\phi)/\omega_d} e(t)\mathrm{d}t + \left[-\int_{(\pi-\phi)/\omega_d}^{(2\pi-\phi)/\omega_d} e(t)\mathrm{d}t + \int_{(2\pi-\phi)/\omega_d}^{(3\pi-\phi)/\omega_d} e(t)\mathrm{d}t \right.$

$\displaystyle\left. -\int_{(3\pi-\phi)/\omega_d}^{(4\pi-\phi)/\omega_d} e(t)\mathrm{d}t + \int_{(4\pi-\phi)/\omega_d}^{(5\pi-\phi)/\omega_d} e(t)\mathrm{d}t - \cdots \right] = \int_0^{(\pi-\phi)/\omega_d} e(t)\mathrm{d}t$

$$-\sum_{k=0}^{\infty}\int_{[(2k+1)\pi-\phi]/\omega_d}^{[(2k+2)\pi-\phi]/\omega_d}e(t)dt + \sum_{k=0}^{\infty}\int_{[(2k+2)\pi-\phi]/\omega_d}^{[(2k+3)\pi-\phi]/\omega_d}e(t)dt. \text{ Note that}$$

$$\int_a^b e(t)dt = \frac{\zeta\sin(\omega_d t+\phi)e^{-\zeta\omega_n t}\big|_b^a}{\omega_n\sqrt{1-\zeta^2}} + \frac{\cos(\omega_d t+\phi)e^{-\zeta\omega_n t}\big|_b^a}{\omega_n}.$$

$$\int_0^{\infty}|e(t)|dt = \frac{2\zeta}{\omega_n} + \frac{2}{\omega_n}\frac{e^{-\frac{\zeta}{\sqrt{1-\zeta^2}}(\pi-\phi)}}{1-e^{-\frac{\zeta}{\sqrt{1-\zeta^2}}\pi}}. \quad \phi = \tan^{-1}\frac{\sqrt{1-\zeta^2}}{\zeta}.$$

[3] $G_{er}(s) = \dfrac{E(s)}{R(s)} = \dfrac{s+1}{s+2} = \dfrac{1+j\omega}{2+j\omega} = \sqrt{\dfrac{1+\omega^2}{4+\omega^2}}e^{j(\tan^{-1}\omega-\tan^{-1}\omega/2)},$

$$\therefore e_{ss} = \sqrt{\frac{1+\omega^2}{4+\omega^2}}\Bigg|_{\omega=1} \sin\left[t+30°+\left(\tan^{-1}\omega-\tan^{-1}\frac{\omega}{2}\right)\Big|_{\omega=1}\right]$$

$$-\sqrt{\frac{1+\omega^2}{4+\omega^2}}\Bigg|_{\omega=2}\cos\left[2t-45°+\left(\tan^{-1}\omega-\tan^{-1}\frac{\omega}{2}\right)\Big|_{\omega=2}\right]$$

$$= 0.632\sin(t+48.43°) - 0.791\cos(2t-26.57°).$$

[4] A. $L(s) = P(s)C(s) = \dfrac{K_1 K_2(s+3)}{s^2(s+1)(s+2)}, \quad K_P = \lim\limits_{s\to 0}L(s) = \infty,$

$\quad K_v = \lim\limits_{s\to 0}sL(s) = \infty, \quad K_a = \lim\limits_{s\to 0}s^2 L(s) = \dfrac{3K_1 K_2}{2}, \quad e_{ss} = \dfrac{0.4}{K_1 K_2}.$

B. $G_{ed}(s) = -\dfrac{P(S)}{1+P(s)C(s)}, \quad D(s) = 1+\dfrac{1}{s}, \quad e_{ss} = \lim\limits_{s\to 0}sG_{ed}(s)D(s) = 0.$

[5] A. $e_{ss} = \dfrac{1}{1+K_P}\cdot\dfrac{1}{1+K_P} = 0.05$ yields $K_P = 19.$

B. $e_{ss} = \dfrac{1+K_P K_F}{1+K_P}. \quad 1+K_P K_F = 0$ yields $K_F = -\dfrac{1}{K_P} = -\dfrac{1}{19}.$

[6] A. $L(s) = \dfrac{50K}{s(s+10)}. \quad K_a = \lim\limits_{s\to 0}s^2 L(s) = 0. \quad e_{ss}(\infty) = \dfrac{1}{K_a} = \infty.$

B. $G_{er}(s) = \dfrac{1-C_F(s)P(s)}{1+KP(s)} = \dfrac{s[Ts^2+(10T+1-50b_2)s+(10-50b_1)]}{Ts^3+(10T+1)s^2+10(1+5KT)s+50K}.$

$\quad e_{ss} = \lim\limits_{s\to 0}sG_{er}(s)\cdot\dfrac{1}{s^3}.$

Choosing $b_1 = \dfrac{1}{5}$ and $b_2 = \dfrac{10T+1}{50}$ ensures $e_{ss} = 0.$

★ Chapter 7

Basic Level

[1] $y(t) = K\left(1 - e^{-t/T}\right)$. $y(t_{0.1}) = 0.1K$ gives $K\left(1 - e^{-t_{0.1}/T}\right) = 0.1K$.
Thus, $t_{0.1} = -T\ln 0.9$. In the same manner, $t_{0.9} = -T\ln 0.1$.
$\therefore T_r = t_{0.9} - t_{0.1} = 2.2T$.
$y(t_s) = 0.95K$ yields $K\left(1 - e^{-t_s/T}\right) = 0.95K$. Thus, $T_s = 3T$.

[2] (1) $K = 0.1$: gm $= 21.3\,\text{dB}$ and PM $= 67.5°$; $K = 0.2$: gm $= 15.3\,\text{dB}$ and PM $= 50.7°$; $K = 0.4$: gm $= 9.3\,\text{dB}$ and PM $= 30.3°$; $K = 0.6$: gm $= 5.8\,\text{dB}$ and PM $= 18.3°$.

(2) A, B, and C: true (0.67, 0.51, 0.30); D: false.

(3) T_r becomes smaller and M_p and T_s become larger as K becomes larger.

[3] A. Substitution $M_p = 15\%$ into (6.24) yields $\zeta = 0.52$. The characteristic polynomial is $D(s) = s^2 + 7s + K$. Thus, $2 \times 0.52\omega_n = 7$ and $\omega_n^2 = K$.
$\therefore K = 45.2$.

B. $e_{ss}(\infty) = \dfrac{7}{K} = 0.155$.

[4] The characteristic polynomial is $D(s) = s^3 + 16s^2 + 51s + 36 + K$. The desired characteristic polynomial is $\phi(s) = (s^2 + 2 \times 0.8\omega_n s + \omega_n^2)(s + p)$.
Thus, $p + 1.6\omega_n = 16$, $1.6p\omega_n + \omega_n^2 = 51$, and $p\omega_n^2 = 36 + K$.
$\therefore K = 30.1$, $\omega_n = 2.32\,\text{rad/s}$, and $p = 12.3$.

[5] A. $\omega_1 = \dfrac{1.8}{T_r} = 12$, $\zeta_1 = 0.60$, $K_1 = \mp1.32$, $\sigma_1 = 6$. Choosing poles to be $-15 \pm j5$ yields $10 + 50K_P T_d = 30$ and $50K_P = 250$. Thus, $K_P = 5$ and $T_d = 0.08$.

B. Choosing poles to be $-15 \pm j5$ and -50 yields $10 + 50K_P T_d = 80$, $50K_P = 1\,750$, and $\dfrac{50K_P}{T_i} = 12\,500$. Thus, $K_P = 35$, $T_i = 0.14$, and $T_d = 0.04$.

[6] A. $C(s) = 1$: $1.34\,\text{dB}$ and $4.1°$, PD: 9.9 dB and $29.8°$.

B. Shorten T_r and reduce M_P and T_s.

[7] A. The characteristic equation is $s^2 + (10 + 500T_d)s + 500 = 0$. Since $\omega = \sqrt{500}\,\text{rad/s}$ and $2\omega\zeta = 10 + 500T_d$, $T_d = \dfrac{2\zeta\omega - 10}{500} = 0.0432$.

B. Shorten T_r and increase M_P.

[8] A. Insert a PI controller $C(s) = K_P \left(1 + \dfrac{1}{T_i s} \right)$ to make the open-loop transfer function of the system a Type-2 one. The characteristic equation is $s^4 + 9s^3 + 18s^2 + KK_P s + \dfrac{KK_P}{T_i} = 0$. The first column of the Routh table is $1 \quad 9 \quad 18 - \dfrac{KK_P}{9} \quad \dfrac{KK_P}{162 - KK_P} \left(162 - KK_P - \dfrac{81}{T_i} \right) \dfrac{KK_P}{T_i}$.

Thus, $0 < KK_P < 162$, $T_i > 0$, and $(162 - KK_P)T_i - 81 > 0$. For selected K_P and T_i, if $K_P > 0$ then $0 < K < \dfrac{162T_i - 81}{K_P T_i}$, and if $K_P < 0$ then $\dfrac{162T_i - 81}{K_P T_i} < K < 0$.

B. Omitted.

[9] A. $E(s) = R(s) - Y(s) = \dfrac{1 - K_F P(s)}{1 + K_P P(s)} R(s) = \dfrac{(0.1s+1)(0.5s+1) - K_F}{(0.1s+1)(0.5s+1) + K_P} R(s)$.

The characteristic polynomial is $(0.1s+1)(0.5s+1) + K_P = 0.05s^2 + 0.6s + 1 + K_P$. $\therefore K_P > -1$.

B. $e_{ss} = \dfrac{1}{1 + K_P}$.

C. $e_{ss} = \dfrac{1 - K_F}{1 + K_P}$ yields $K_F = 1$.

Advanced Level

[1] A. $\bar{P}(s) = \dfrac{2(1 - 0.1s)}{(1 + 0.1s)(s + 2)}$.

B. Omitted.

C. $D_P(s) = 2 + 1.2s + 0.1s^2$ and $N_P(s) = 2 - 0.2s$. $h_0 = 1$, $h_1 = 0.7$, $h_2 = 0.12$, $0.1\sigma^2 - 0.35\sigma + 0.12 = 0$. $\sigma = 0.39$, $K_P = 1.32$, $K_i = 2.60$, $C(s) = K_P + \dfrac{K_i}{s}$.

D. $-3.71 \pm j3.60$ and -1.94.

E. Omitted.

[2] The characteristic polynomial of the inner loop is $s^2 + 17s + 70 + K_f$. Thus, $2 \times 0.7\omega_i = 17$. $K_f = \omega_i^2 - 70 = 77$. The characteristic polynomial of the whole system is $s^3 + 17s^2 + 147s + K$. And it is also $(s^2 + 2 \times 0.5\omega_n s + \omega_n^2)(s + p)$. Thus, $\omega_n + p = 17$, $(p + \omega_n)\omega_n = 147$, and $\omega_n^2 p = K$. $\therefore \omega_n = 8.65\,\mathrm{rad/s}$, $p = 8.35$, and $K = 625$.

★ Chapter 8

Basic Level

[1] High-precision input-output tracking, reject disturbances, suppress effects caused by parameter changes and uncertainties, faster responses.

[2] Two in the right-half and one in the left-half s plane.

[3] Controllable.

[4] A. $|M_C| = |[B\ AB\ A^2B]| = \begin{vmatrix} 1 & 0 & -2-a \\ 0 & -2 & 6 \\ 0 & a & -2+a \end{vmatrix} = 4 - 8a.$

∴ $a \neq 0.5$.

B. $|M_O| = \left| \begin{bmatrix} C \\ CA \\ CA^2 \end{bmatrix} \right| = \begin{vmatrix} 0 & 0 & 1 \\ a & 1 & 1 \\ -2+a & a-2 & -a+1 \end{vmatrix} = a^2 - 3a + 2.$

∴ $a \neq 1$ and $a \neq 2$.

C. $|sI - A| = \begin{vmatrix} s & -1 & 1 \\ 2 & s+3 & 0 \\ -a & -1 & s-1 \end{vmatrix} = s^3 + 2s^2 + (a-1)s + 3a - 4.$

Routh's stability criterion gives $\frac{3}{4} < a < 2$.

D. Controllable and observable: $G(s) = \dfrac{as + 3a - 2}{s^3 + 2s^2 + (a-1)s + 3a - 4}$,

uncontrollable but observable: $G(s) = \dfrac{0.5}{s^2 + 3s + 2.5}$, and controllable

but unobservable: $G(s) = \dfrac{1}{s^2 + s - 1}$ ($a = 1$) and $G(s) = \dfrac{2}{s^2 + 1}$ ($a = 2$).

[5] When $u_r(t)$ is used,

$$M_{Cr} = \begin{bmatrix} B_r & AB_r & A^2B_r & A^3B_r \end{bmatrix} = \begin{bmatrix} 0 & 1 & 0 & -\omega^2 \\ 1 & 0 & -\omega^2 & 0 \\ 0 & 0 & -2\omega & 0 \\ 0 & -2\omega & 0 & 2\omega^3 \end{bmatrix},$$

$|M_{Cr}| = 0$. Uncontrollable.

When $u_\theta(t)$ is used,

$$M_{C\theta} = \begin{bmatrix} B_\theta & AB_\theta & A^2B_\theta & A^3B_\theta \end{bmatrix} = \begin{bmatrix} 0 & 0 & 2\omega & 0 \\ 0 & 2\omega & 0 & -2\omega^3 \\ 0 & 1 & 0 & -4\omega^2 \\ 1 & 0 & -4\omega^2 & 0 \end{bmatrix},$$

$|M_{C\theta}| = -12\omega^4 \neq 0$. Controllable.

When both $u_r(t)$ and $u_\theta(t)$ are used,

$$M_C = \begin{bmatrix} B & AB & A^2B & A^3B \end{bmatrix}$$

$$= \begin{bmatrix} 0 & 0 & 1 & 0 & 0 & 2\omega & -\omega^2 & 0 \\ 1 & 0 & 0 & 2\omega & -\omega^2 & 0 & 0 & -2\omega^3 \\ 0 & 0 & 0 & 1 & -2\omega & 0 & 0 & -4\omega^2 \\ 0 & 1 & -2\omega & 0 & 0 & -4\omega^2 & 2\omega^3 & 0 \end{bmatrix},$$

rank$M_C = 4$. controllable.

[6] $|sI - (A - BF)| = s^2 + (3 + f_2)s_1 + (2 + f_1)$. Routh's stability criterion yeilds that the system is unstable if $f_1 \leq -2$ or $f_2 \leq -3$.

[7] A. $A = \begin{bmatrix} 0 & 1 \\ \dfrac{mgl}{J} & -\dfrac{b}{J} \end{bmatrix}$, $B = \begin{bmatrix} 0 \\ \dfrac{1}{J} \end{bmatrix}$, $C = \begin{bmatrix} 1 & 0 \end{bmatrix}$, $F = \begin{bmatrix} 21.45 & 4.99 \end{bmatrix}$.

   B. $F = \begin{bmatrix} 4.20 & 1.75 \end{bmatrix}$.

   C. and D. Omitted.

[8] Controllable canonical form: $A = \begin{bmatrix} 0 & 1 \\ 0 & -10 \end{bmatrix}$, $B = \begin{bmatrix} 0 \\ 1 \end{bmatrix}$, $C = \begin{bmatrix} 50 & 0 \end{bmatrix}$.

   Observable canonical form: $A = \begin{bmatrix} 0 & 0 \\ 1 & -10 \end{bmatrix}$, $B = \begin{bmatrix} 50 \\ 0 \end{bmatrix}$, $C = \begin{bmatrix} 0 & 1 \end{bmatrix}$.

[9] A. $P(s) = \dfrac{s+2}{s^2 + 14s + 45}$. $A = \begin{bmatrix} 0 & 1 \\ -45 & -14 \end{bmatrix}$, $B = \begin{bmatrix} 0 \\ 1 \end{bmatrix}$, and $C = \begin{bmatrix} 2 & 1 \end{bmatrix}$.

   B. Let $L = \begin{bmatrix} l_1 \\ l_2 \end{bmatrix}$. $|sI - (A - LC)| = \begin{vmatrix} s + 2l_1 & l_1 - 1 \\ 2l_2 + 45 & s + l_2 + 14 \end{vmatrix} = s^2 + $
   $(2l_1 + l_2 + 14)s + (2l_2 - 17l_1 + 45)$. The desired characteristic polynomial is $\phi(s) = s^2 + 144s + 14\,400$. $|sI - (A - LC)| = \phi(s)$ provides $l_1 = -671.2$ and $l_2 = 1\,472.4$.

[10] Designing a state-feedback gain for the dual system (A^T, C^T) and transposing it yields $L = \begin{bmatrix} 30 & 200 \end{bmatrix}^T$.

[11] Substitution $M_p = 10\%$ into (6.24) yields $\zeta_1 = 0.60$. Substitution ζ_1 and

T_s $(= 0.1)$ into (6.29) yields $\omega_1 = 48\,\mathrm{rad/s}$. Choosing dominant poles to be $-50 \pm j40$ and another pole to be -150 yields $L = \begin{bmatrix} 226 & 15\,067 & 462\,621 \end{bmatrix}^{\mathrm{T}}$.

【12】 A. $P(s) = \dfrac{10}{s^3 + 15s^2 + 50s}$. Controllable canonical form:

$$A = \begin{bmatrix} 0 & 1 & 0 \\ 0 & 0 & 1 \\ 0 & -50 & -15 \end{bmatrix}, \; B = \begin{bmatrix} 0 \\ 0 \\ 1 \end{bmatrix}, \; C = \begin{bmatrix} 10 & 0 & 0 \end{bmatrix}.$$

B. $Q = \mathrm{diag}\{100, 100, 1\}$ $R = 1$ yields $L = \begin{bmatrix} 10.02 & 1.74 & -4.97 \end{bmatrix}^{\mathrm{T}}$.

C. $Q = \mathrm{diag}\{100, 1, 1\}$ $R = 1$ yields $F = \begin{bmatrix} 10 & 2.97 & 0.23 \end{bmatrix}$.

D. Omitted.

Advanced Level

【1】 A. (1) $M_C = \begin{bmatrix} B & AB & A^2B \end{bmatrix} = \begin{bmatrix} 1 & 0 & 0 \\ 0 & 1 & a \\ c & bc & b^2c \end{bmatrix}$.

$|M_C| = (b - a)bc \neq 0$ yields $a \neq b$, $b \neq 0$, and $c \neq 0$.

(2) $M_O = \begin{bmatrix} C \\ CA \\ CA^2 \end{bmatrix} = \begin{bmatrix} 0 & 1 & d \\ 1 & a & bd \\ a & a^2 & b^2d \end{bmatrix}$.

$|M_O| = (a - b)bd \neq 0$ yields $a \neq b$, $b \neq 0$, and $d \neq 0$.

Controllable but not observable: $a \neq b$, $b \neq 0$, $c \neq 0$, and $d = 0$.
Observable but not controllable: $a \neq b$, $b \neq 0$, $c = 0$, and $d \neq 0$.
Controllable and observable: $a \neq b$, $b \neq 0$, $c \neq 0$, and $d \neq 0$.

B. Controllable and observable: $G(s) = C(sI - A)^{-1}B = \dfrac{cds^2 + (1 - acd)s - b}{s(s - a)(s - b)}$.

Controllable but not observable: $G(s) = \dfrac{1}{s(s - a)}$.

Observable but not controllable: $G(s) = \dfrac{1}{s(s - a)}$.

C. Poles: $s(s - a)(s - b) = 0$. A pole at $s = 0$.

$\therefore$ not asymptotically stable.

【2】 A. $\bar{P}(s) = \dfrac{e^{-0.01s}}{s(s + 10)} = \dfrac{s^2 - 600s + 120\,000}{s^4 + 610s^3 + 126\,000s^2 + 1\,200\,000s}$

B. $A = \begin{bmatrix} 0 & 1 & 0 & 0 \\ 0 & 0 & 1 & 0 \\ 0 & 0 & 0 & 1 \\ 0 & -1\,200\,000 & -126\,000 & -610 \end{bmatrix}$, $B = \begin{bmatrix} 0 \\ 0 \\ 0 \\ 1 \end{bmatrix}$,

$C = \begin{bmatrix} 120\,000 & -600 & 1 \end{bmatrix}$.

C. Let $Q_L = 100 \times \mathrm{diag}\{1,1,1,1\}$ and $R_L = 1$.

$L = \begin{bmatrix} 10 & 0.52 & -5.23 & -47.7 \end{bmatrix}^{\mathrm{T}}$.

D. $Q = \mathrm{diag}\{100, 100, 1, 1\}$ and $R = 1$. $F = \begin{bmatrix} 10 & 1.05 & 0.013 & 0.00084 \end{bmatrix}$.

E. Omitted.

★ Chapter 9

Basic Level

[1] C.   **[2]** B.   **[3]** C.

[4] A. $G_{\omega u}(s) = \dfrac{K_T}{(L_a s + R_a)(Js + b) + K_T K_E}$.

Letting $b = 0$ gives $T_e T_m s^2 + T_m s + 1 = 0$.

B. Solving $T_e T_m s^2 + T_m s + 1 = 0$ and letting $s_2 \leq 5 s_1$ yields $T_e \leq$
$\dfrac{[1 - (2/3)^2] T_m}{4} = 0.14 T_m$.

C. $T_e = \dfrac{L_a}{R_a} = 0.000769\,\mathrm{s}$ and $T_m = \dfrac{J R_a}{K_T K_E} = 0.00900\,\mathrm{s}$.

$T_e < 0.14 T_m$. $G_{\omega u}(s) = \dfrac{K_T}{(L_a s + R_a)(Js + b) + K_T K_E}$ and a reduced-

order model is $G_{\omega u}^r(s) = \dfrac{K_T}{R_a(Js + b) + K_T K_E}$.

D. $T_e = \dfrac{L_a}{R_a} = 0.05\,\mathrm{s}$ and $T_m = \dfrac{J R_a}{K_T K_E} = 0.900\,\mathrm{s}$. $T_e < 0.14 T_m$ is not
true.

[5] $C(z) = \dfrac{1.25 \times 10^{-5}(z + 0.997)}{z^2 - 1.99z + 0.99}$.

[6] PD: $C(z) = \dfrac{52.1(z - 0.989)}{z - 0.882}$. PID: $C(z) = \dfrac{346.2(z^2 - 1.977z + 0.977)}{(z - 1)(z - 0.778)}$.

Advanced Level

[1] A. $G_P(s) = \dfrac{2.25 e^{-0.0017s}}{0.012 s^2 + 0.097 s + 1}$.

B. $\omega_c = 15\,\text{rad/s}.$ $\dfrac{0.1}{\omega_c} = 0.00667 \gg 0.0017.$ (9.4) holds and T_D can be ignored.

C. The selected poles $\{-20 \pm j10, -100\}$ yield $8.083 + 2.25K_P T_d = 140$, $83.333 + 2.25K_P = 4\,500$, and $\dfrac{2.25K_P}{T_i} = 50\,000.$ Thus, $K_P = 23.556$, $T_d = 0.0299$, and $T_i = 0.0883$.

D. Omitted.

[2] A. Omitted.

B. $C(s) = 1 + \dfrac{1}{s}$. Euler's method: $C(z) = 1 + \dfrac{h}{z-1}$.

C. Backward difference: $C(z) = 1 + \dfrac{hz}{z-1}$.

D. Tustin transform: $C(z) = 1 + \dfrac{h/2(z+1)}{z-1}$.

E. and F. Omitted.

★ Chapter 10

⟨ **Basic Level** ⟩

[1] A.   **[2]** D.

[3] Comparing $(s^2 + \sqrt{2}\omega_n s + \omega_n^2)(s+p) = 0$ and $(s^2 + 2s + 0.25)(s+2) + K = 0$ yields $\sqrt{2}\omega_n + p = 4$, $\omega_n^2 + \sqrt{2}p\omega_n = 4.25$, and $p\omega_n^2 = 0.5 + K$. Thus, $\omega_n = 0.892$, $K = 1.679$, and $p = 2.739$. Closed-loop poles: $-0.631 \pm 0.63j$ and -2.739. Closed-loop zeros: -2.

[4] Riccati equation is $2P - 2P^2 + 2 = 0$. $P = \dfrac{(1+\sqrt{5})}{2}$. $F = R^{-1}B^T P = 1 + \sqrt{5}$. $A - BF = -\sqrt{5}$. Stable.

[5] $\omega_1 = \dfrac{1.8}{T_r} = 18$, $\zeta_1 = 0.6$, $K_1 = \mp 1.3$, $\sigma_1 = 6$. Choosing poles to be $-20 \pm j10$ and -100 yields $F_R = -12\,500$ and $F_P = \begin{bmatrix} -8\,007 & 131 \end{bmatrix}$.

[6] A. $K_P = 75.4$, $T_i = 0.1$, $T_d = 0.0335$.

B. (10.6) holds.

C.∼F. Omitted.

[7] A. Controllable and observable.

B. $A_R = 0$ and $B_R = 1$. $\bar{A} = \begin{bmatrix} A_R & -B_R C \\ 0 & A \end{bmatrix}$, $\bar{B} = \begin{bmatrix} 0 \\ B \end{bmatrix}$, and $\bar{C} = \begin{bmatrix} 0 & C \end{bmatrix}$.

C. Controllable and observable.

D. $L = \begin{bmatrix} 972.3 & -11.5 & -0.111 \end{bmatrix}^{\text{T}}$.

E. $F_R = -1$, $F = \begin{bmatrix} 0.396 & 0.0458 & -0.437 \end{bmatrix}$.

F. Omitted.

Advanced Level

[1] A. $E(s) = \dfrac{(1 + K_1 K_2)s + 3 - K_2}{(1 + K_1 K_2)s + 3 + K_1 K_2} \dfrac{1}{s}$.

B. $e_{ss} = \dfrac{3 - K_2}{3 + K_1 K_2}$.

C. $\left. \dfrac{\partial e_{ss}}{\partial K_1} \right|_{\substack{K_1=100 \\ K_2=0.1}} = -0.00172$ and $\left. \dfrac{\partial e_{ss}}{\partial K_2} \right|_{\substack{K_1=100 \\ K_2=0.1}} = -1.793$.

[2] A. For $Q_K = \text{diag}\{100, 1, 1\}$ and $R_K = 1$, $\bar{F} = \begin{bmatrix} -10 & 1.42 & 18.30 \end{bmatrix}$.

B. For the observer poles $0.5 \times \{-4.128 + j4.591 \quad -4.128 - j4.591\}$,

$L = \begin{bmatrix} 10.91 & -0.94 \end{bmatrix}$; $1 \times \{-4.128 + j4.591 \quad -4.128 - j4.591\}$,

$L = \begin{bmatrix} 8.89 & -0.28 \end{bmatrix}$; $2 \times \{-4.128 + j4.591 \quad -4.128 - j4.591\}$,

$L = \begin{bmatrix} 13.98 & 1.04 \end{bmatrix}$; $5 \times \{-4.128 + j4.591 \quad -4.128 - j4.591\}$,

$L = \begin{bmatrix} 102.42 & 5.00 \end{bmatrix}$; and $10 \times \{-4.128 + j4.591 \quad -4.128 - j4.591\}$,

$L = \begin{bmatrix} 493.79 & 11.61 \end{bmatrix}$;

[3] A.

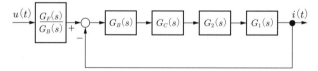

Figure A10.1   Solution of Advanced-level Problem 3 A

B. $T_F = 0.0018\,\text{s}$.

C. $P(s) = G_B(s)G_2(s)G_1(S)$.

D. $P(s) = \dfrac{205}{s + 5} - \dfrac{447}{s + 303} + \dfrac{242}{s + 556} \doteq \dfrac{205}{s + 5} - \dfrac{447}{303} + \dfrac{242}{556} = \dfrac{205}{s + 5} - 1.04$.

E. $\omega_1 = 1.2$, $\zeta_1 = 0.7$, $K_1 = \mp 1.02$, $\sigma_1 = 1.5$. Choosing the poles to be $\{-5 + j2 \quad -5 - j2\}$ yields $K_P = 0.0256$ and $T_i = 0.181$.

F. Omitted.

Index

—— 著者略歴 ——

佘　錦華（しゃ　きんか）
1993年　東京工業大学大学院理工学研究科博士
　　　　課程修了（制御工学専攻），博士（工学）
1993年　東京工科大学講師
2001年　東京工科大学助教授
2007年　東京工科大学准教授
2010年　東京工科大学教授
　　　　現在に至る

禹　珍碩（う　じんそく）
2017年　首都大学東京大学院システムデザイン
　　　　研究科博士後期課程修了（システムデ
　　　　ザイン専攻ヒューマンメカトロニクス
　　　　システム学域），博士（工学）
2017年　首都大学東京特任助教
2019年　東京工科大学助教
2022年　東京工科大学講師
　　　　現在に至る

大山　恭弘（おおやま　やすひろ）
1985年　東京工業大学大学院理工学研究科博士
　　　　課程修了（制御工学専攻），工学博士
1991年　東京工科大学講師
1996年　東京工科大学助教授
2002年　東京工科大学教授
　　　　現在に至る

英語で学ぶ　**制御システム設計**
Introduction to Linear Control System Design
© Jinhua She, Jinseok Woo, Yasuhiro Ohyama

2022 年 4 月 15 日　初版第 1 刷発行　　　　　　　　　　★

検印省略

著　者　　佘　　　錦　　　華
　　　　　禹　　　珍　　　碩
　　　　　大　山　恭　弘
発 行 者　株式会社　コロナ社
　　　　　代表者　牛来真也
印 刷 所　三美印刷株式会社
製 本 所　有限会社　愛千製本所

112–0011　東京都文京区千石 4–46–10
発 行 所　株式会社　コロナ社
CORONA PUBLISHING CO., LTD.
Tokyo Japan
振替 00140–8–14844・電話(03)3941–3131(代)
ホームページ https://www.coronasha.co.jp

ISBN 978–4–339–03238–3　C3053　Printed in Japan　　　（松岡）

JCOPY　＜出版者著作権管理機構 委託出版物＞
本書の無断複製は著作権法上での例外を除き禁じられています。複製される場合は，そのつど事前に，
出版者著作権管理機構（電話 03-5244-5088, FAX 03-5244-5089, e-mail: info@jcopy.or.jp）の許諾を
得てください。

本書のコピー，スキャン，デジタル化等の無断複製・転載は著作権法上での例外を除き禁じられています。
購入者以外の第三者による本書の電子データ化及び電子書籍化は，いかなる場合も認めていません。
落丁・乱丁はお取替えいたします。

システム制御工学シリーズ

(各巻A5判，欠番は品切です)

■編集委員長 池田雅夫
■編集委員 足立修一・梶原宏之・杉江俊治・藤田政之

定価は本体価格+税です。
定価は変更されることがありますのでご了承下さい。

図書目録進呈◆

計測・制御テクノロジーシリーズ

（各巻A5判，欠番は品切または未発行です）

■計測自動制御学会 編

定価は本体価格＋税です。
定価は変更されることがありますのでご了承下さい。

||||||||||||||||||||||||||||||||||||||||| 図書目録進呈◆

機械系教科書シリーズ

(各巻A5判，欠番は品切です)

- ■編集委員長　木本恭司
- ■幹　　事　平井三友
- ■編集委員　青木　繁・阪部俊也・丸茂榮佑

配本順	書名	著者	頁	本体
1.（12回）	機械工学概論	木本恭司 編著	236	2800円
2.（1回）	機械系の電気工学	深野あづさ 著	188	2400円
3.（20回）	機械工作法（増補）	平井三友・和田任弘・塚本晃久 共著	208	2500円
4.（3回）	機械設計法	三田純義・朝比奈奎一・黒田孝二 共著	264	3400円
5.（4回）	システム工学	古川正志・荒井誠・吉浜克己 共著	216	2700円
6.（5回）	材料学	久保井徳洋・樫原恵蔵 共著	218	2600円
7.（6回）	問題解決のための Cプログラミング	佐藤次男・中村理一郎 共著	218	2600円
8.（32回）	計測工学（改訂版）—新SI対応—	前田良一・田村昭郎・押田至啓 共著	220	2700円
9.（8回）	機械系の工業英語	牧野州秀・生水雅之 共著	210	2500円
10.（10回）	機械系の電子回路	高橋晴俊・阪部雄也 共著	184	2300円
11.（9回）	工業熱力学	丸茂榮佑・木本恭司 共著	254	3000円
12.（11回）	数値計算法	藪忠司・伊藤悼 共著	170	2200円
13.（13回）	熱エネルギー・環境保全の工学	木本民恭司・山﨑友紀雄彦 共著	240	2900円
15.（15回）	流体の力学	坂本雅彦・坂田光雄 共著	208	2500円
16.（16回）	精密加工学	田口紘一・明石剛二 共著	200	2400円
17.（30回）	工業力学（改訂版）	吉村靖夫・米内山誠 共著	240	2800円
18.（31回）	機械力学（増補）	青木繁 著	204	2400円
19.（29回）	材料力学（改訂版）	中島正貴 著	216	2700円
20.（21回）	熱機関工学	越智敏明・老固潔光・吉本隆也 共著	206	2600円
21.（22回）	自動制御	阪部俊一・飯田賢一 共著	176	2300円
22.（23回）	ロボット工学	早川恭弘・欒川明彦・矢野順一 共著	208	2600円
23.（24回）	機構学	重松洋男・大高一 共著	202	2600円
24.（25回）	流体機械工学	小池勝 著	172	2300円
25.（26回）	伝熱工学	丸茂榮佑・矢尾匡永・牧野秀州 共著	232	3000円
26.（27回）	材料強度学	境田彰芳 編著	200	2600円
27.（28回）	生産工学 —ものづくりマネジメント工学—	本位田光重・皆川健多郎 共著	176	2300円
28.（33回）	CAD／CAM	望月達也 著	224	2900円

定価は本体価格+税です。
定価は変更されることがありますのでご了承下さい。

図書目録進呈◆

ロボティクスシリーズ

（各巻A5判，欠番は品切です）

- ■編集委員長　有本　卓
- ■幹　　　事　川村貞夫
- ■編集委員　石井　明・手嶋教之・渡部　透

定価は本体価格＋税です。
定価は変更されることがありますのでご了承下さい。

図書目録進呈◆